Monique Rothe

Gut microbiota & nutrition

Monique Rothe

# Gut microbiota & nutrition

Response of Escherichia coli to dietary factors in the murine intestine

Südwestdeutscher Verlag für Hochschulschriften

**Impressum / Imprint**
Bibliografische Information der Deutschen Nationalbibliothek: Die Deutsche Nationalbibliothek verzeichnet diese Publikation in der Deutschen Nationalbibliografie; detaillierte bibliografische Daten sind im Internet über http://dnb.d-nb.de abrufbar.
Alle in diesem Buch genannten Marken und Produktnamen unterliegen warenzeichen-, marken- oder patentrechtlichem Schutz bzw. sind Warenzeichen oder eingetragene Warenzeichen der jeweiligen Inhaber. Die Wiedergabe von Marken, Produktnamen, Gebrauchsnamen, Handelsnamen, Warenbezeichnungen u.s.w. in diesem Werk berechtigt auch ohne besondere Kennzeichnung nicht zu der Annahme, dass solche Namen im Sinne der Warenzeichen- und Markenschutzgesetzgebung als frei zu betrachten wären und daher von jedermann benutzt werden dürften.

Bibliographic information published by the Deutsche Nationalbibliothek: The Deutsche Nationalbibliothek lists this publication in the Deutsche Nationalbibliografie; detailed bibliographic data are available in the Internet at http://dnb.d-nb.de.
Any brand names and product names mentioned in this book are subject to trademark, brand or patent protection and are trademarks or registered trademarks of their respective holders. The use of brand names, product names, common names, trade names, product descriptions etc. even without a particular marking in this works is in no way to be construed to mean that such names may be regarded as unrestricted in respect of trademark and brand protection legislation and could thus be used by anyone.

Coverbild / Cover image: www.ingimage.com

Verlag / Publisher:
Südwestdeutscher Verlag für Hochschulschriften
ist ein Imprint der / is a trademark of
OmniScriptum GmbH & Co. KG
Heinrich-Böcking-Str. 6-8, 66121 Saarbrücken, Deutschland / Germany
Email: info@svh-verlag.de

Herstellung: siehe letzte Seite /
Printed at: see last page
**ISBN: 978-3-8381-3798-8**

Zugl. / Approved by: Potsdam, Universität Potsdam, Mathematisch-Naturwissenschaftliche Fakultät, Dissertation, 2013

Copyright © 2014 OmniScriptum GmbH & Co. KG
Alle Rechte vorbehalten. / All rights reserved. Saarbrücken 2014

## ABSTRACT

Diet is a major force influencing the intestinal microbiota. This is obvious from drastic changes in microbiota composition after a dietary alteration. Due to the complexity of the commensal microbiota and the high inter-individual variability, little is known about the bacterial response at the cellular level. The objective of this work was to identify mechanisms that enable gut bacteria to adapt to dietary factors. For this purpose, germ-free mice monoassociated with the commensal *Escherichia coli* K-12 strain MG1655 were fed three different diets over three weeks: a diet rich in starch, a diet rich in non-digestible lactose and a diet rich in casein. Two-dimensional gel electrophoresis and electrospray-tandem mass spectrometry were applied to identify differentially expressed proteins of *E. coli* recovered from small intestine and caecum of mice fed the lactose or casein diets in comparison with those of mice fed the starch diet. Selected differentially expressed bacterial proteins were characterised *in vitro* for their possible roles in bacterial adaptation to the various diets. Proteins belonging to the oxidative stress regulon *oxyR* such as the alkyl hydroperoxide reductase subunit F (AhpF), the DNA protection during starvation protein (Dps) and the ferric uptake regulatory protein (Fur), which are required for *E. coli*'s oxidative stress response, were upregulated in *E. coli* of mice fed the lactose-rich diet. Reporter gene analysis revealed that not only oxidative stress but also carbohydrate-induced osmotic stress led to the OxyR-dependent expression of *ahpCF* and *dps*. Moreover, the growth of *E. coli* mutants lacking the *ahpCF* or *oxyR* genes was impaired in the presence of non-digestible sucrose. This indicates that some OxyR-dependent proteins are crucial for the adaptation of *E. coli* to osmotic stress conditions. In addition, the function of two so far poorly characterised *E. coli* proteins was analysed: the 2-deoxy-D-gluconate 3-dehydrogenase

(KduD) was upregulated in intestinal *E. coli* of mice fed the lactose-rich diet and this enzyme and the 5-keto 4-deoxyuronate isomerase (KduI) were downregulated on the casein-rich diet. Reporter gene analysis identified galacturonate and glucuronate as inducers of the *kduD* and *kduI* gene expression. KduI was shown to facilitate the breakdown of these hexuronates, which are normally degraded by the uronate isomerase (UxaC), the altronate oxidoreductase (UxaB), the altronate dehydratase (UxaA), the mannonate oxidoreductase (UxuB) and the mannonate dehydratase (UxuA), whose expression was repressed by osmotic stress. The growth of *kduID*-deficient *E. coli* on galacturonate or glucuronate was impaired in the presence of osmotic stress, suggesting KduI and KduD to compensate for the function of the regular hexuronate degrading enzymes under such conditions. This indicates a novel function of KduI and KduD in *E. coli*'s hexuronate metabolism. Promotion of the intracellular formation of hexuronates by lactose connects these *in vitro* observations with the induction of KduD on the lactose-rich diet. Taken together, this study demonstrates the crucial influence of osmotic stress on the gene expression of *E. coli* enzymes involved in stress response and metabolic processes. Therefore, the adaptation to diet-induced osmotic stress is a possible key factor for bacterial colonisation of the intestinal environment.

# TABLE OF CONTENTS

ABSTRACT .................................................................................................. 1
LIST OF FIGURES ....................................................................................... 6
LIST OF TABLES ......................................................................................... 7
ABBREVIATIONS ........................................................................................ 9
1. INTRODUCTION .................................................................................. 13
 1.1. Intestinal microbiota ......................................................................... 13
 1.2. Effects of intestinal colonisation – bacteria-host interaction ............ 13
 1.3. Nutrient exchange between host and microbiota ............................ 15
 1.4. Microbiota composition – the impact of diet ................................... 17
 1.5. Gnotobiotic animal models ............................................................. 18
 1.6. Model organism *Escherichia coli* .................................................. 19
 1.7. Adaptation of *E. coli* to the intestinal environment ....................... 20
2. OBJECTIVES ....................................................................................... 23
3. MATERIAL AND METHODS ............................................................... 24
 3.1. Animal experiment .......................................................................... 24
  3.1.1. Gnotobiotic mouse experiment ................................................. 24
  3.1.2. Preparation of intestinal contents ............................................. 25
  3.1.3. 16S rRNA gene profile of intestinal bacteria ............................ 26
  3.1.4. Preparation of proteins for two-dimensional gel electrophoresis .... 27
  3.1.5. Labelling of bacterial proteins .................................................. 28
  3.1.6. First dimension ......................................................................... 29
  3.1.7. Second dimension .................................................................... 29
  3.1.8. Detection of differentially expressed proteins .......................... 29
  3.1.9. Identification of differentially expressed proteins ..................... 30
  3.1.10. Determination of dietary substrates in the luminal gut contents ..... 31
 3.2. *In vitro* analysis ............................................................................ 33
  3.2.1. Generation of luciferase reporter gene constructs ................... 33
  3.2.2. Luciferase reporter gene assays .............................................. 35

  3.2.2.1. *ahpCF* and *dps* promoters ............................................................. 35
  3.2.2.2. *kduI* and *kduD* promoters ............................................................. 36
  3.2.2.3. *uxaC*, *uxaB* and *uxuAB* promoters .............................................. 37
 3.2.3. Generation of knock-out mutants ........................................................ 38
 3.2.4. Complementation of knock-out mutants ............................................. 38
 3.2.5. Growth experiments ............................................................................ 40
  3.2.5.1. Growth of *E. coli* strains Δ*ahpCF* and Δ*oxyR* .............................. 40
  3.2.5.2. Growth of *E. coli* strain Δ*kduID* .................................................... 41
  3.2.5.3. Growth of complemented *E. coli* strains ......................................... 42
 3.2.6. Determination of medium osmolality .................................................... 42
 3.2.7. Determination of intracellular reactive oxygen species ..................... 42
 3.2.8. Overexpression of KduI and KduD ..................................................... 43
 3.2.9. Determination of hexuronate conversion in the presence of KduI and KduD ................................................................................... 45
 3.2.10. Determination of intracellular hexuronate concentration ................... 47
3.3. Statistical analysis ........................................................................................ 47
**4.** **RESULTS** ........................................................................................................... 49
4.1. Characterisation of the gnotobiotic mouse model ........................................ 49
 4.1.1. Cell numbers of intestinal *E. coli* ...................................................... 49
 4.1.2. 16S rRNA gene sequencing profile ................................................... 49
 4.1.3. Intestinal substrate availability ........................................................... 50
4.2. Bacterial adaptation to the host diets ........................................................... 52
 4.2.1. Identification of differentially expressed *E. coli* proteins .................. 52
 4.2.2. Proof of principle: Adaptation of intestinal *E. coli* to the host diets ..................................................................................................... 54
 4.2.3. Induction of stress-related proteins on the lactose diet ..................... 56
 4.2.4. Regulation of the uncharacterised *E. coli* proteins KduI and KduD ..................................................................................................... 57
4.3. *In vitro* characterisation of selected proteins ............................................... 59
 4.3.1. Induction of OxyR-dependent proteins by a lactose-rich diet ........... 59

    4.3.1.1. Induction of the oxyR regulon by osmolytes ........................... 59
    4.3.1.2. Positive correlation of *ahpCF* and *dps* expression and medium osmolality................................................................. 62
    4.3.1.3. No formation of intracellular $H_2O_2$ by osmotic stress ............... 64
    4.3.1.4. Requirement of OxyR-regulated proteins at high osmolality .... 65
  4.3.2. Analysis of proteins KduI and KduD ............................................. 67
    4.3.2.1. Induction of *kduI* and *kduD* gene expression by hexuronates ............................................................................ 67
    4.3.2.2. Repression of *uxaCA*, *uxaB* and *uxuAB* expression by osmotic stress ..................................................................... 69
    4.3.2.3. No repression of *kduI* and *kduD* gene expression by osmotic stress ..................................................................... 73
    4.3.2.4. Promotion of hexuronate conversion by KduI and KduD ......... 74
    4.3.2.5. Requirement of KduID for growth on hexuronates at high osmolality ................................................................................ 76
    4.3.2.6. Origin of hexuronates in the mouse intestine ........................... 79
**5. DISCUSSION** ........................................................................................ **82**
5.1. Gnotobiotic mice – a useful model to analyse the effect of dietary factors ........................................................................................... 82
5.2. Adaptation of intestinal *E. coli* to dietary factors ............................. 85
5.3. OxyR-dependent stress-response proteins are crucial under osmotic stress ................................................................................. 86
5.4. OxyR-dependent proteins are also required for other commensal bacteria ............................................................................................ 92
5.5. KduI and KduD are required for hexuronate metabolism at high osmolality ......................................................................................... 93
5.6. Concluding remarks ....................................................................... 101
APPENDIX I: REFERENCES ................................................................... 103
APPENDIX II: SUPPLEMENTAL MATERIAL ............................................ 122

## LIST OF FIGURES

Figure 1. Interaction between host and microbiota ................................... 14
Figure 2. Nutrient digestion and the impact of the gut microbiota ............. 16
Figure 3. Adaptation of *E. coli* to the intestinal environment ................... 22
Figure 4. Principle of pre-labelling ............................................................ 28
Figure 5. Principle of the 2D-DIGE analysis ............................................. 31
Figure 6. Schematic presentation of luciferase reporter gene constructs ... 35
Figure 7. Deleted chromosomal regions in mutant *E. coli* strains ............ 39
Figure 8. Schematic presentation of pGEM-T constructs ......................... 44
Figure 9. Overexpression of KduI and KduD ............................................ 46
Figure 10. Colony forming units (CFU) of intestinal *E. coli* after 3 weeks of feeding mice diets rich in starch, lactose or casein ................ 49
Figure 11. Concentrations of glucose, lactose and amino acids in the gut contents of mice fed the starch, the lactose or the casein diet .... 51
Figure 12. Representative two-dimensional gel images of the proteome of intestinal *E. coli* .................................................................. 53
Figure 13. Functional categories of differentially expressed *E. coli* proteins ..................................................................................... 54
Figure 14. Induction of Leloir pathway enzymes in *E. coli* of mice fed the lactose diet ............................................................................... 55
Figure 15. Regulation of KduD and KduI in intestinal *E. coli* .................... 58
Figure 16. Induction of the *ahpCF* and *dps* promoters by osmolytes ........... 60
Figure 17. OxyR-dependency of osmolyte-induced *ahpCF* and *dps* expression ................................................................................ 61
Figure 18. Positive correlation of *ahpCF* and *dps* promoter activity and medium osmolality .................................................................. 63
Figure 19. No formation of intracellular $H_2O_2$ by osmotic stress ............... 65
Figure 20. Repression of the *uxaCA*, *uxaB* and *uxuAB* promoters by osmolytes ................................................................................ 71

Figure 21. OxyR-dependency of the *uxaCA*, *uxaB* and *uxuAB* gene expression ...... 72

Figure 22. Effect of osmotic stress on hexuronate-induced expression of *kduI* and *kduD* ...... 74

Figure 23. Breakdown of galacturonate and glucuronate by KduI and KduD ...... 75

Figure 24. Growth retardation of *E. coli* Δ*kduID* by osmotic stress ...... 78

Figure 25. Generation of intracellular hexuronate in exponential phase of *E. coli* during growth on lactose ...... 81

Figure 26. Potential mechanisms of OxyR activation by osmotic stress in *E. coli* ...... 90

Figure 27. Role of KduI and KduD in *E. chrysanthemi* and *E. coli* ...... 96

Figure 28. Potential mechanism of how osmotic stress may influence hexuronate degrading enzymes ...... 99

Figure I. Schematic representation of pKEST-MR ...... 122

Figure II. Representative 16S rRNA gene sequence of *E. coli* ...... 123

Figure III. Osmolality of the various media used for the luciferase reporter gene assays ...... 125

## LIST OF TABLES

Table 1. Composition of the semisynthetic diets fed to the gnotobiotic mice for 3 weeks ...... 25

Table 2. Regulation of bacterial proteins belonging to cell redox homeostasis and stress-response processes with expression change ≥ 2-fold ($P < 0.05$) . ...... 56

Table 3. Growth of wild type, *ahpCF* and *oxyR* deficient *E. coli* in the presence or absence of osmotic stress caused by non-fermentable sucrose ...... 66

| | | |
|---|---|---|
| Table 4. | Growth of *ahpCF* and *oxyR* deficient *E. coli* containing complementing plasmids with the corresponding genes and promoters in comparison to the wild type with the empty vector under aerobic conditions | 67 |
| Table 5. | Induction of the *kduD* and *kduI* promoters in *E. coli* by the various mouse diets, small intestinal mucosal tissues, carbohydrates, casamino acids or hexuronates under aerobic and anaerobic growth conditions | 68 |
| Table 6. | Growth behaviour of wild type and *kduID* deficient *E. coli* on galacturonate- or glucuronate-containing medium with or without osmotic stress caused by non-fermentable sucrose | 77 |
| Table 7. | Growth behaviour of *kduID* deficient *E. coli* containing complementing plasmids in comparison to the wild type with the empty vector under aerobic conditions | 79 |
| Table 8. | Concentrations of galacturonate and glucuronate in the intestinal contents of mice fed the starch, the lactose or the casein diet determined with an uronate dehydrogenase assay. | 80 |
| Table I. | Primers used for generation of luciferase reporter gene constructs, deletion mutants, complementing plasmids and pGEM-T-Easy vectors | 126 |
| Table II. | Identified proteins of *E. coli* obtained from small intestine and caecum of mice fed the starch, the lactose or the casein diet with differential expression factors of ≥ 2 | 127 |
| Table III. | Specific activities of KduI and KduD calculated for hexuronate concentrations observed after incubation of cell-free extracts of *E. coli* clones overexpressing KduI, KduD or both with 10 mM galacturonate or 10 mM glucuronate at 37°C | 131 |

## ABBREVIATIONS

| | |
|---|---|
| 1-Way AOV | one-way analysis of variance |
| 16S rRNA | 16S ribosomal RNA of the prokaryotic small ribosomal subunit |
| AhpC | alkyl hydroperoxide reductase subunit C |
| AhpF | alkyl hydroperoxide reductase subunit F |
| AhpR | alkyl hydroperoxide reductase |
| *B. thetaiotamicron* | *Bacteroides thetaiotamicron* |
| bp | base pairs |
| CarA | carbamoyl-phosphate synthase |
| CHAPS | 3-[(3-cholamidopropyl)-dimethylammonio]-1-propanesulfonate |
| Cae | caecum |
| $ddH_2O$ | double distilled water |
| CFU | colony forming unit |
| $CO_2$ | carbon dioxide |
| Cy | fluorescent cyanine minimal dyes |
| 2D-DIGE | two-dimensional difference gel electrophoresis |
| DMSO | dimethylsulfoxid |
| DNA | desoxyribonukleinsäure |
| Dps | DNA protection during starvation protein |
| DTT | dithiothreitol |
| EB | equilibration buffer |
| *E. coli* | *Escherichia coli* |
| EF | elongation factor |
| FabE | acetyl-CoA carboxylase |
| FabI | enoyl-[acyl-carrier-protein] reductase |
| FNR | DNA-binding transcriptional dual regulator FNR |
| Fur | ferric uptake regulatory protein |

| | |
|---|---|
| × g | centrifugal force (as gravity) |
| GalE | UDP-glucose 4-epimerase |
| GalK | galactokinase |
| GalM | galactose mutarotase |
| GalT | galactose-1-phosphate uridylyltransferase |
| GalU | uridylyltransferase |
| GapA | glyceraldehyde-3-phosphate dehydrogenase |
| GdhA | glutamate dehydrogenase |
| GlsA1 | glutaminase 1 |
| GltX | glutamyl-tRNA synthetase |
| GorB | quinone oxidoreductase 2 |
| GlyA | hydroxymethyltransferase |
| GntT | gluconate transporter |
| GrcA | autonomous glycyl radical cofactor |
| $H_2$ | hydrogen |
| Gst | glutathione S transferase |
| HisS | histidyl-tRNA synthetase |
| HslU | ATP dependent protease ATPase subunit HslU |
| HybC | hydrogenase 2 large chain |
| IPG strip | immobilised pH gradient strip |
| IPTG | isopropyl-β-D-thiogalactopyranosid |
| KatG | catalase peroxidase |
| Kbl | the 2-amino-3-ketobutyrate coenzyme A ligase |
| KduD | 2-deoxy-D-gluconate 3-dehydrogenase |
| KduI | 5-keto 4-deoxyuronate isomerase |
| $K^+$ | potassium |
| LacY | lactose permease |
| lacZ | β-galactosidase |
| LB | lysogeny broth |
| $Log_{10}$ | logarithm to the base of 10 |

| | |
|---|---|
| Lux | bacterial luciferase |
| NaCl | sodium chloride |
| NAD | nicotinamide adenine dinucleotide |
| NADH | nicotinamide adenine dinucleotide hydrogen |
| NfnB | oxygen insensitive NAD(P)H nitroreductase |
| NusA | transcription elongation protein nusA |
| $OD_{600}$ | optical density at 600 nm |
| Omp | outer membrane protein |
| OxyR | OxyR DNA-binding transcriptional dual regulator |
| PAGE | polyacrylamide gel electrophoresis |
| PBS | phosphate buffered saline |
| PCR | polymerase chain reaction |
| PgK | phosphoglycerate kinase |
| Pgm | phosphoglucomutase |
| Pnp | polyribonucleotide nucleotidyltransferase |
| PurC | phosphoribosylaminoimidazole-succinocarboxamide synthase |
| PurE | $N^5$-carboxyaminoimidazole ribonucleotide mutase |
| PurH | AICAR transformylase |
| Pyk | pyruvate kinase |
| PyrBI | aspartate carbamoyltransferase |
| RBS | ribosomal binding site |
| rpm | rotation rate (rounds per minute) |
| Rps | 30S ribosomal protein |
| SCFA | short chain fatty acid |
| SDS | sodium dodecyl sulphate |
| SOC | Super Optimal broth with Catabolite repression |
| SodB | superoxide dismutase (Fe) |
| SI | small intestine |
| TB | Terrific Broth |

| | |
|---|---|
| TCA | trichloroacetic acid |
| $t_d$ | doubling time |
| Tpx | thiol peroxidase |
| Tris | tris(hydroxymethyl)aminomethane |
| TrxB | thioredoxin reductase |
| Udh | uronate dehydrogenase |
| UDP | uridinphosphate |
| Ugd | UDP-glucose 6-dehydrogenase |
| UxaA | altronate dehydratase |
| UxaB | altronate oxidoreductase |
| UxaC | uronate isomerase |
| UxuA | mannonate dehydratase |
| UxuB | mannonate oxidoreductase |
| U-test | Mann-Whitney-U-test |
| Vh | volt hours |
| vs. | versus |
| w/v | weight per volume |
| wt/wt | weigt per weight |
| X-Gal | 5-bromo-4-chloro-indolyl-β-D-galactopyranoside |
| µ | specific growth rate |

An "s" behind an abbreviation indicates the plural.

# 1. INTRODUCTION

## 1.1. Intestinal microbiota

The gastrointestinal tract of mammals harbours a complex microbial community comprising 10-times more bacterial cells than eukaryotic host cells [for review see: BENGMARK, 1998] and 150-times more bacterial than host genes [QIN, 2010]. The microbiota reaches approximately $10^{12}$ cells per g of faeces (dry mass) [FRANKS, 1998; SIMMERING, 1999] and consists of 1,000–1,150 bacterial species [QIN, 2010], the majority of which have not yet been cultivated *in vitro*. Although the gut microbiota displays high diversity at the species and subspecies level, only low diversity was observed at the phylum level [GILL, 2006; TURNBAUGH, 2009a; TURNBAUGH, 2010]. Approximately 95% of the entire microbiota belongs to the major bacterial phyla Firmicutes (including gram-positive genera such as *Clostridium*, *Eubacterium*, *Ruminococcus*, *Butyrivibrio*, *Anaerostipes*, *Roseburia*, *Faecalibacterium*), Bacteroidetes (including the gram-positive genus *Bifidobacterium*), Actinobacteria and Proteobacteria (including the gram-negative *Enterobacteriaceae* with *Escherichia coli* as the most prominent representative) [TAP, 2009; QIN, 2010]. However, the high inter-individual variability in intestinal microbiota composition hampers the definition of a core microbiome present in everyone [QIN, 2010].

## 1.2. Effects of intestinal colonisation – bacteria-host interaction

Despite the differences in species composition, many functions exerted by the intestinal microbiota are apparently shared among individuals. Known functions of the gut microbiota that affect host health include (Figure 1): i) Salvage of energy by conversion of non-digestible carbohydrates such as dietary fibre to short chain fatty acids (SCFA) [for review see: SAVAGE, 1986; GIBSON, 2004]. ii) Provision of a barrier against pathogenic bacteria by the use of available habitats and metabolic substrates [for review see: VAN DER WAAIJ, 1989; ARANEO, 1996]. iii) Contribution to the maturation and maintenance of the immune system [BRY, 1996; SLACK, 2009; LEE, 2010b; for review see: ROUND, 2009]. iv) Regulation of the differentiation and growth of the intestinal epithelium [for review see: Falk, 1998]. v) Activation of secondary plant metabolites such as polyphenols, which are supposed to have anti-allergic, anti-inflammatory and anti-carcinogenic effects [DE SOUSA, 2007; for review see: YAMAMOTO, 2001].

However, the mechanisms how host factors, such as health status and nutrition, influence the composition and the activity of the microbiota are poorly understood. Most investigations of bacteria-host interactions regard the host response or the activity of pathogenic bacteria [for review see: ADLERBERTH, 2000].

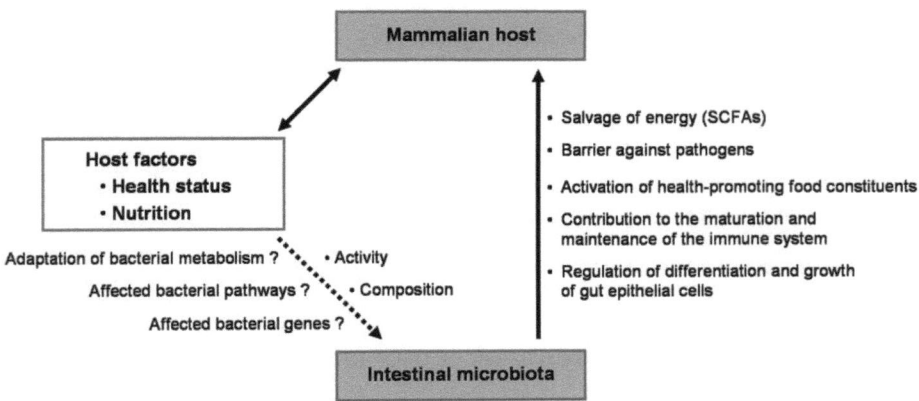

Figure 1.    Interaction between host and microbiota. The microbial community inhabiting the human intestinal tract largely affects host physiology, while host factors, such as health status and nutrition, influence the composition and the activity of the intestinal microbiota. However, the mechanisms underlying the adaptation of commensal bacteria to host factors are poorly understood. SCFAs, short chain fatty acids.

## 1.3. Nutrient exchange between host and microbiota

Extraction of energy from otherwise indigestible food compounds is one of the main functions of the intestinal microbiota. Therefore, diet provides nutrients not only for the host but also for the indigenous bacteria. In the small intestine, diet-derived low-molecular weight nutrients, such as monosaccharides and amino acids, are directly absorbed by the host [for review see: FERRARIS, 2001], whereas disaccharides, such as sucrose and maltose, polysaccharides, such as starches, as well as peptides and complex proteins are digested by various host enzymes and subsequently absorbed by specific transporters.

However, mammals have limited capacities for the hydrolysation of complex non-starch polysaccharides originating from plant material, such as cellulose and pectin [for review see: HOOPER, 2002]. Together with nutrients that escape host digestion and absorption, indigestible dietary fibres are degraded by a large variety of bacterial polysaccharidases, glycosidases, proteases and peptidases to smaller oligomers and their component carbohydrates and amino acids [for review see: MACKIE, 1997]. These low-molecular substances are fermented by the indigenous microbiota to SCFAs (acetic, propionic, butyric acids), hydroxy and dicarboxylic organic acids, $H_2$, $CO_2$ and other neutral, acidic or basic end products (Figure 2).

In the large intestine, SCFAs are rapidly absorbed and metabolised by various host tissues [for review see: SAVAGE, 1986; GIBSON, 2004]. Theoretical calculations and comparisons of the energy consumption of conventional microbiota associated mice and germ-free mice revealed that SCFAs contribute 10 to 30% to daily energy requirement of the host [WOSTMANN, 1983; for review see: MCNEIL, 1984; MACKIE, 1997]. Accordingly, these fermentation products are potential energy sources and provide a metabolic benefit to the host.

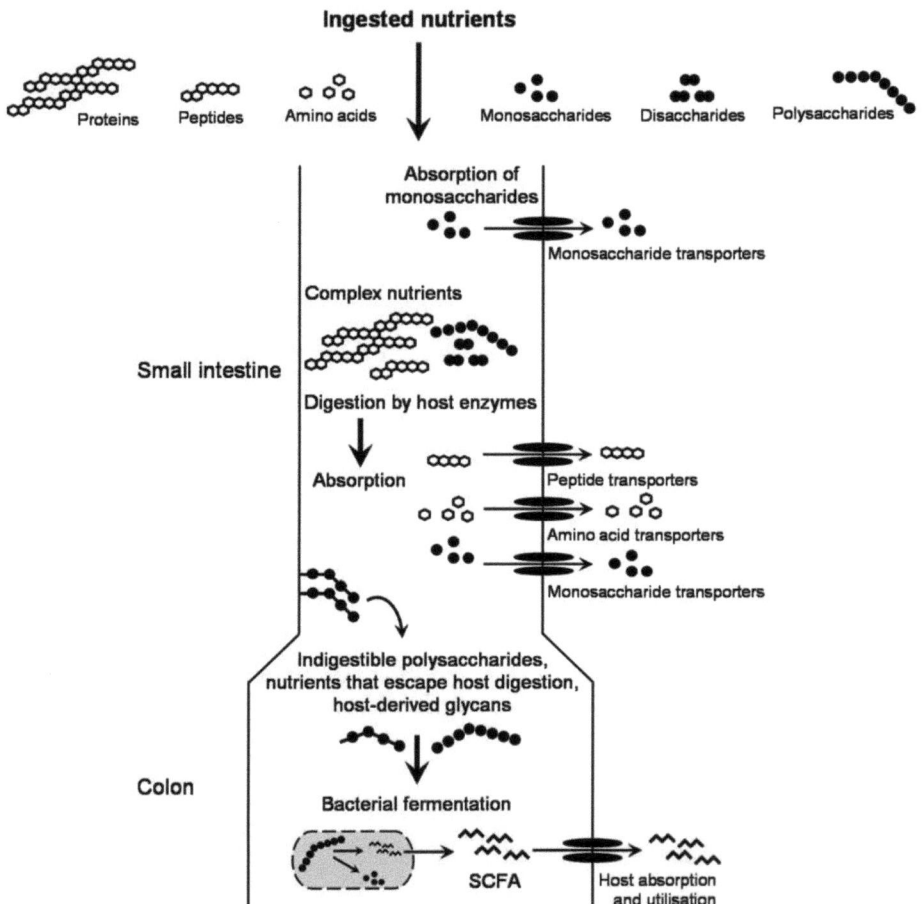

**Figure 2.    Nutrient digestion and the impact of the gut microbiota.** Simple carbohydrates and amino acids are absorbed by active transport in the proximal region of the small intestine. Proteins, peptides and polysaccharides are digested by host enzymes and subsequently absorbed by specific transporters. Nutrients that escape host digestion, indigestible carbohydrates, such as dietary fibres, and host-derived nutrients, such as glycans, are fermented by the intestinal microbiota and bacterial end-products, such as short chain fatty acids (SCFA), are absorbed and utilised by the host [modified after HOOPER, 2002].

## 1.4. Microbiota composition – the impact of diet

Since dietary nutrients are the main carbon and energy sources for intestinal bacteria, diet is one of the major determinants for the persistence of a given species in the gastrointestinal tract. Intestinal bacteria capable of taking advantage of a given dietary nutrient proliferate faster than those unable to do so. This is obvious from 13-fold higher numbers of faecal anaerobes in rats fed a high-fibre diet compared to rats fed a diet devoid of dietary fibre [MACZULAK, 1993]. Indigestible oligofructose has been demonstrated to stimulate the growth of bifidobacteria in the intestine of human subjects and rodents [GIBSON, 1995; KLEESSEN, 2001].

However, not only the type of dietary fibre but also the macronutrient composition of diet has been shown to influence the composition and the activity of the gut microbiota as reflected by alterations in intestinal microbiota composition after weaning [FAVIER, 2002] or after a dietary switch [LEY, 2008; Turnbaugh, 2009b; FAITH, 2011; WU, 2011; PATRONE, 2012]. Studies in mice colonised with human feacal microbial communities demonstrated a rapid change of the microbiota composition following a switch from a low-fat, dietary fibre-rich diet to a high-fat, high-sugar Western-style diet. Moreover, the gene expression of enzymes catalysing the conversion of amino acids, nucleotide sugars and carbohydrates changed rapidly in response to the dietary shifts [Turnbaugh, 2009b]. The analysis of the human gut microbiome revealed an increase in genes encoding bacterial enzymes involved in amino acid degradation, simple carbohydrate catabolism, vitamin biosynthesis and bile salt metabolism in response to a protein-rich host diet [YATSUNENKO, 2012].

Taken together, these studies not only revealed that dietary alterations affect the intestinal microbiota composition and thereby indirectly gastrointestinal function and host health but also that the microbial changes are accompanied by modulations in the metabolic activity at the cellular level. Although the metabolic response of representative bacterial species to simple parameters, such as substrate availability, pH, osmolality and oxygen partial pressure, has extensively been studied *in vitro*, the mechanisms underlying bacterial adaptation to host, environmental and dietary factors are poorly understood.

## 1.5. Gnotobiotic animal models

Studying the effects of dietary alterations at the cellular level is hampered by the complexity of conventional microbial communities and the high inter-individual variability [DENOU, 2009]. Complex model systems such as human microbiota associated mice or streptomycin-treated mice, in which gram-negative but not gram-positive bacteria were eliminated [MYHAL, 1982], cannot exclude that diet-induced alterations at the population level are responsible for metabolic changes observed at the cellular level. To circumvent this problem, simplified model systems are required.

Gnotobiotic animals are a useful tool to analyse bacterial colonisation under defined and controlled environmental conditions [FALK, 1998; HOOPER, 2002; HOOPER, 2012]. Since the effects of the resident microbiota can be ruled out, this model is an excellent tool to investigate and define molecular mechanisms of host-microbe interactions [BUTTERTON, 1996; KAMIYA, 1997]. One well-known example of a gnotobiotic animal model are mice monoassociated with *Bacteroides thetaiotamicron*, a dominant representative of the human distal small intestinal microbiota [MOORE, 1974]. Using this model, *B. thetaiotaomicron* was shown to restore the synthesis of fucosylated glycoconjugates in the intestinal mucosa throughout adulthood whereas in germ-free mice production ceases after weaning. *B. thetaiotaomicron* produces specific enzymes for the metabolism of these host-derived oligosaccharides enabling this bacterium to persistently colonise the murine gut [BRY, 1996; HOOPER, 1999]. This study clearly illustrates that bacteria are capable of adapting to the intestinal environment and that they trigger the host to provide a growth substrate. Moreover, this example demonstrates that gnotobiotic animals are exellent tools to investigate the molecular mechanisms underlying the bacterial adaptation to host-endogenous factors.

## 1.6. Model organism *Escherichia coli*

To analyse the molecular mechanisms of bacterial adaptation to the intestinal environment, it is worthwhile to use a model organism that is a representative of the entire microbiota and easily cultivable *in vitro*. The organism used in the actual study is the gram-negative, rod-shaped *Escherichia coli*. This bacterium is the most abundant facultatively anaerobic bacterium present in the intestine of humans [ESCHERICH, 1988] and most mammals [MANSSON, 1957; CLAPPER, 1963; DUOBOS, 1963; PESTI, 1963]. Although not numerically dominant, this species was shown to prevent the establishment of potentially pathogenic bacteria in the gut of rodents [HUDAULT, 2001].

*E. coli* is able to efficiently adapt and change its metabolism in response to its environment. This bacterium proliferate under aerobic, microaerobic and anoxic conditions, in which mixed-acid fermentation results in the production of lactate, succinate, formate, ethanol, acetate and $CO_2$ [INGLEDEW, 1984]. Its ability to switch between aerobic and anaerobic respiration in dependence of the oxygen availability allows an easy cultivation of *E. coli in vitro* and enables this bacterium to successfully colonise the intestine of germ-free rodents [for review see: ESPEY, 2013].

In the last decades, laboratory strains of *E. coli* have been investigated intensively in various genetic, biochemical and physiological studies. Therefore, *E. coli* is one of the best examined gut microbes. The genome of the laboratory strain MG1655 was sequenced completely [BLATTNER, 1997], enabling precise functional, transcriptional, phenotypical and biochemical analysis of mutant strains or expressed proteins. These molecular techniques

## 1.7. Adaptation of *E. coli* to the intestinal environment

In the mammalian intestine, *E. coli* colonises the polysaccharide-rich outer mucus layer [MÖLLER, 2003; for review see: KIM, 2010], which is generated by the host's intestinal epithelium and composed of glycoproteins, glycolipids and epithelial cell debris [PEEKHAUS, 1998]. The *E. coli* genome encodes approximately 40 glycoside hydrolases [for review see: HENRISSAT, 1997], but no enzymes for the degradation of complex polysaccharides [HOSKINS, 1985]. Therefore, in the intestine, only mono- and disaccharides as well as short malto-oligosaccharides released from diet- or host-derived polysaccharides are potential energy sources for this bacterium [PEEKHAUS, 1998; CHANG, 2004]. Complex polysaccharides are primarily degraded by polysaccharide hydrolase enzymes secreted by the dominant anaerobic microbiota [COMSTOCK, 2003] or by colonic epithelial cells of the host [BEAULIEU, 1991].

Competition for nutrients is a key factor for bacterial colonisation of the mammalian intestine. FRETER *et al.* postulated that the concentration of preferred nutrients affects the population size of the bacterial species in the intestine [FRETER, 1983]. Since *E. coli* co-metabolise a large variety of only transiently available nutrients, which in addition are present at low concentrations, this organism has an enhanced ability to successfully colonise the mammalian intestine [JONES, 2007; FABICH, 2008].

Not only the adaptation to changing nutrient availability but also the flexibility of the aerobic respiratory system and the use of the best available electron acceptors enhance the colonisation efficiency of intestinal *E. coli* (Figure 3). Mouse colonisation experiments with wild type and mutant *E. coli* lacking genes encoding respiratory oxidases and reductases, such as *cydAB*, *cydDC*, *fnr* and *arcA*, demonstrated that this organism is dependent on both microaerobic and anaerobic respiration to successfully compete with other organisms in the mouse gut. This promotes the view that the intestinal tract of mammals is not completely anoxic [JONES, 2007]. Oxygen-rich tissues surrounding the gut lumen promote oxygen diffusion into the intestine at levels relevant for bacterial metabolism [HE, 1999; for review see: ESPEY, 2013]. The intestinal oxygen concentration is of major importance for intestinal colonisation because it has been demonstrated that oxygen-intolerant bacteria failed to colonise the intestine of germ-free rodents unless preceded by inoculation with either aerobic or facultative aerobic microorganisms [for review see: ESPEY, 2013].

Colonisation of the mouse intestine requires in addition the ability to synthesise nucleotides as indicated by a study of VOGEL-SCHEEL et al., who observed a higher expression of key enzymes of the purine and pyrimidine biosynthesis in intestinal compared to *in vitro* grown *E. coli*. Mutants of *E. coli* devoid of the corresponding genes (*pyrBI, purC, ydjG, nanA*) were washed out of the intestinal tract or decreased by several orders of magnitude when inoculated with their respective wild type into germ-free mice [VOGEL-SCHEEL, 2010].

These observations clearly indicate that effective survival *in vivo* requires metabolic flexibility (Figure 3). As indicated by the results of VOGEL-SCHEEL *et al.*, the analysis of the whole *E. coli* proteome obtained from the intestines of gnotobiotic mice is an appropriate method to determine the mechanisms underlying bacterial adaptation to the host environment. Nevertheless, this simplified and defined model has so far not been used to identify molecular mechanisms underlying the adaptation of intestinal bacteria to dietary factors. Due to the large influence of diet on basic parameters, such as intestinal motility, pH, osmolality and nutrient availability, it is of crucial interest to analyse how intestinal bacteria adapt to these factors *in vivo*. Such analysis may contribute to the identification of mechanisms underlying the observed changes of the intestinal microbiota composition in response to dietary factors.

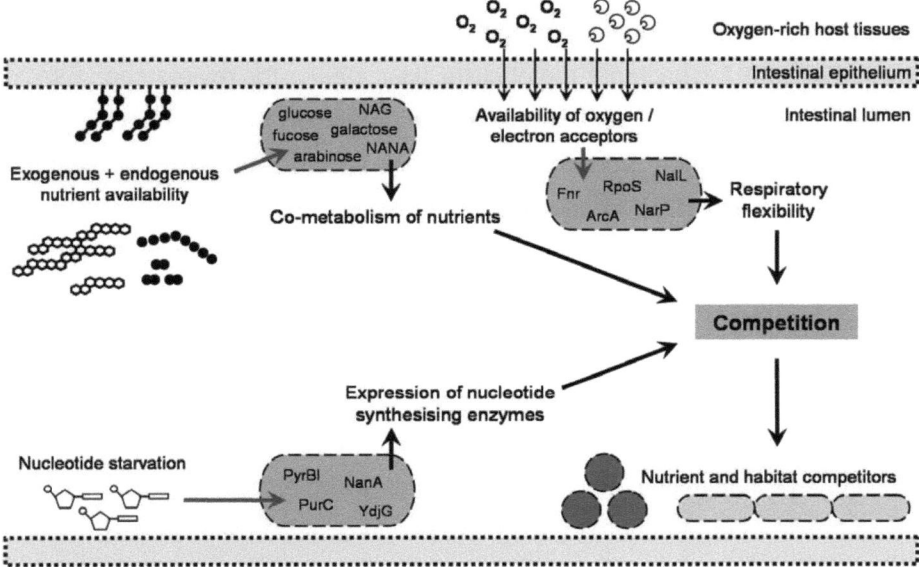

**Figure 3. Adaptation of *E. coli* to the intestinal environment.** Intestinal *E. coli* adapt their metabolism to the *in vivo* conditions by i) simultaneous utilisation of a variety of different energy sources, ii) induction of enzymes involved in purine and pyrimidine biosynthesis and iii) activation of respiratory pathways in response to the availability of oxygen and electron acceptors. This flexibility enables *E. coli* to compete with other intestinal bacteria and to successfully colonise the mammalian intestine.

## 2. OBJECTIVES

Since diet is a major factor influencing the intestinal microbiota, it is important to understand the molecular mechanisms underlying diet-induced bacterial adaptation. However, the focus of the majority of investigations in this field deals with the host-response to bacterial colonisation while the adaptation of commensal bacteria to dietary factors is poorly understood. Therefore, this project aimed to identify mechanisms that enable commensal bacteria to adapt to dietary factors in the mammalian intestine.

The following questions were addressed:

- Does intestinal *E. coli* adapt its metabolism to different host diets?

- Does the *E. coli* proteome reflect the cellular adaptation to intestinal substrate availability?

- Do intestinal parameters other than nutrient availability result in metabolic responses in intestinal *E. coli*?

- What are the underlying molecular mechanisms of bacterial adaptation?

## 3. MATERIAL AND METHODS

### 3.1. Animal experiment

#### 3.1.1. Gnotobiotic mouse experiment

This project aimed to identify mechanisms that enable intestinal bacteria to adapt to dietary factors in the mouse intestine. Therefore, gnotobiotic mice monoassociated with the commensal *E. coli* K-12 strain MG1655 were fed either one of three different diets: a diet rich in starch, a diet rich in non-digestible lactose and a diet rich in casein. For this purpose, three groups of germfree C3H mice (n = 18 to 21; Charles River Laboratories, Wilmington, USA), 9 to 12 weeks of age, were housed in sterile Trexler-type isolators (Metall & Plastik, Radolfzell, Germany) at constant room temperature of 22°C ± 10%, air humidity of 55% ± 10% and a light/dark cycle of 12 h. All materials were sterilised by irradiation (50 kGy) or autoclaving (121°C, 15 min) and transferred to the isolators via a tank filled with 10% chloramine. One week prior to the association with *E. coli*, the mice were switched from standard chow (Altromin 1310 standard, Altromin Spezialfutter GmbH & Co. KG, Lage, Germany) to one of the experimental diets (starch diet, lactose diet, casein diet; Table 1). All diets were sterilised by irradiation at 25 kGy. The germfree status of the mice was checked prior the experiment by Gram stains and cultivation of faecal samples under aerobic and anaerobic growth conditions on Thioglycolate-broth (SIFIN, Berlin, Germany) and Wilkins-Chalgren broth (Oxoid, Hampshire, United Kingdom) [KAMLAGE, 1999]. For monoassociation of mice, *E. coli* K-12 strain MG1655 was cultured anaerobically until mid exponential phase at 37°C in LB-Lennox medium (Carl Roth, Karlsruhe, Germany) and washed in phosphate buffered saline (PBS; $Na_2HPO_4$ [80 g/litre], KCl [2 g/litre], $Na_2HPO_4$ [14.4 g/litre], $KH_2PO_4$ [2.4 g/litre]; pH 7.4). Each mouse was inoculated once with $1 \times 10^7$ bacterial cells in 100 µl PBS by gastric gavage. Mice were killed 21 days after monoassociation by cervical dislocation and intestinal contents of small intestine, caecum and colon were collected.

Table 1. Composition of the semisynthetic diets fed to the gnotobiotic mice for 3 weeks.

| Substrate | % (wt/wt) in indicated diet | | |
|---|---|---|---|
| | Starch diet | Lactose diet | Casein diet |
| Sucrose | 20 | 20 | 20 |
| Starch | 43 | 33 | 3 |
| Lactose | 0 | 10 | 0 |
| Casein | 20 | 20 | 60 |
| Cellulose | 5 | 5 | 5 |
| Sunflower oil | 5 | 5 | 5 |
| Vitamins | 2 | 2 | 2 |
| Minerals | 5 | 5 | 5 |

## 3.1.2. Preparation of intestinal contents

Intestinal contents were weighed and diluted 1:10 (wt/vol) with PBS containing a protease inhibitor mix (1:100 dilution of a 100 × stock; GE Healthcare, Munich, Germany) and cooled on ice. Homogenisation was done by agitation with a Uniprep 24 (speed 2; Uniequip, Planegg, Germany) in the presence of glass beads with a diameter of 2.85 to 3.33 mm. Samples were centrifuged at 300 × g and 4°C for 3 min to eliminate coarse particles. Cell numbers of *E. coli* were determined by viable cell counts. Therefore, supernatants were serially diluted in PBS and aliquots of 100 µl were plated on LB-Lennox agar (Roth, Karlsruhe, Germany). Bacterial cell counts are expressed as log10 per g dry weight of intestinal material. Dry weight was determined by lyophilisation of coarse particles (Loc-1m; Martin Christ Gefriertrocknungsanlagen, Osterode, Germany).

To collect the bacterial cells, supernatants were centrifuged at 10,000 × g and 4°C for 3 min. Bacterial pellets were resuspended in washing buffer (10 mM Tris, pH 8; 5 mM magnesium acetate; 30 µg/ml chloramphenicol;

100 × protease inhibitor mix, 1:100 diluted). Remaining particles originating from the diet were removed by Nycodenz (Axis-shield PoC, Oslo, Norway) gradient centrifugation. For this purpose, 0.5 ml of the cell suspension were pipetted on 0.5 ml Nycodenz solution (40% [wt/vol]) and centrifuged at 186,000 × g and 4°C for 15 min. *E. coli* cells were collected from the interphase and washed four times with 1 ml washing buffer (centrifugation at 10,000 × g and 4°C for 3 min). Isolated *E. coli* cells were stored at -80°C.

### 3.1.3. 16S rRNA gene profile of intestinal bacteria

To exclude contaminations, DNA of representative caecal samples was isolated (RTP Bacteria DNA Mini Kit, Invitek, Berlin, Germany) and bacterial 16S rRNA genes were amplified by polymerase chain reaction (PCR) with the primers 27-f (5'-AGA GTT TGA TCC TGG CTC AG-3') and 1492-r (5'-TAC CTT GTT ACG ACT T-3') [KAGEYAMA, 1999]. PCR was performed in a 24-µl reaction mixture including 1.2 µl of each primer (10 µM, Eurofins MWG Operon, Ebersberg, Germany), 0.38 µl 12.5 mM dNTP (Invitek, Berlin, Germany), 2.4 µl 10 × Dream Taq Green buffer, 1.25 U Dream Taq DNA polymerase (Fermentas, St. Leon-Rot, Germany), and 10 ng template DNA. PCR conditions were as follows: denaturation for 4 min at 94°C; 25 cycles of 1 min at 94°C, 1 min at 55°C, and 1 min at 72°C; and a final extension step of 10 min at 72°C. The size of the PCR products was controlled in 1% agarose gels (Serva, Heidelberg, Germany) with the help of the GeneRuler™ 100 bp Plus DNA Ladder (Fermentas, St. Leon-Rot, Germany). Bands were cut out of the agarose gels and purified (innuPREP Gel Extraction Kit, Analytik Jena, Jena, Germany). DNA concentration was determined using a NanoDrop ND-1000 spectrophotometer (Peqlab, Erlangen, Germany). Amplicons were commercially sequenced using the 1492-r primer and the cycle sequencing technology (Eurofins MWG Operon, Ebersberg, Germany).

## 3.1.4. Preparation of proteins for two-dimensional gel electrophoresis

All steps were performed on ice. Frozen bacterial pellets were thawed on ice, resuspended in 800 µl lysis buffer (8 M urea, 30 mM Tris, 4% [wt/vol] 3-[(3-cholamidopropyl)-dimethylammonio]-1-propanesulfonate (CHAPS); pH 8.5) and incubated on ice for 5 min. Resuspended samples were pipetted into 1.5 ml micro tubes (Sarstedt, Nümbrecht, Germany) that contained 1.2 g of 0.1 mm zirconia-silica beads (Roth, Karlsruhe, Germany). Cell disruption was performed in a FP120 FastPrep cell disruptor (Thermo Scientific, Waltham, MA, USA) at a speed of 4.0 m/s and a run time of 20 s. Three cycles were interrupted by incubation of samples on ice for 5 min. To collect the protein extracts, the bottom of the micro tubes was tapped with a sterile needle. Micro tubes were inserted into 2 ml Eppendorf tubes and centrifuged at 2,000 × g for 2 min. To eliminate unbroken cells, samples were centrifuged at 14,000 × g and 4°C for 20 min. Supernatants were collected and bacterial DNA was removed by incubation with 125 U Benzonase (Novagen, Merck KGaA, Darmstadt, Germany) at 37°C for 5 min. Isolated proteins in the supernatants were enriched and purified from components interfering with the two-dimensional difference gel electrophoresis (2D-DIGE) by selective precipitation of proteins (2-D clean-up kit; GE Healthcare, Munich, Germany) according to the manufacturer's instructions. Purified protein solutions were adjusted to pH 8.5 using 50 mM NaOH to ensure optimal binding of the fluorescent dyes. Protein concentrations were determined by the method of Bradford [BRADFORD, 1976] with a ready to use reagent mixture (Bio-Rad, Madrid, Spain). Bovine serum albumin was used as a reference protein. To reach the protein concentration, which was needed for 2D-DIGE, samples were pooled.

## 3.1.5. Labelling of bacterial proteins

Bacterial proteins were labelled with fluorescent cyanine minimal dyes (Cy). These dyes have specific excitation wavelengths and are therefore suitable for the separation of three different samples on the same SDS-PAGE gel. This enables the use of an internal standard, which is present on every gel (Figure 4). For each sample, 50 µg purified protein were transferred to Eppendorf tubes and labelled with CyDye DIGE Fluor Cy3 or Cy5 (GE Healthcare, Munich, Germany) as described by the manufacturer. For preparation of the internal standard, 25 µg protein of each sample was pipetted into a separate Eppendorf tube. This mixture was labelled with CyDye DIGE Fluor Cy2 (GE Healthcare, Munich, Germany). To run different samples on the same 2D gel, Cy3- and Cy5-labelled samples were combined with 50 µg of the Cy2-labelled internal standard into a new Eppendorf tube.

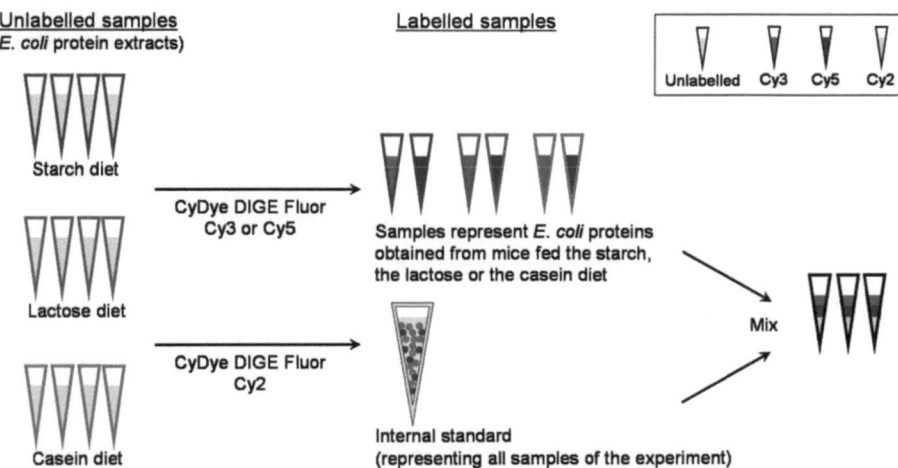

**Figure 4. Principle of pre-labelling.** Purified bacterial proteins obtained from intestinal *E. coli* of mice fed the starch, the lactose or the casein diet were labelled with CyDye DIGE Fluor Cy3 and Cy5. The internal standard was prepared by pooling aliquots of all samples and was labelled with CyDye DIGE Fluor Cy2. To run different samples on the same 2D gel, Cy3- and Cy5-labelled samples were combined with an aliquot of the internal standard into a new Eppendorf tube. Cy, fluorescent cyanine minimal dyes (modified after Ettan DIGE System user manual, GE Healthcare, 2005).

## 3.1.6. First dimension

Labelled samples were prepared for isoelectric focusing as follows: internal standard, sample A, sample B (50 µg protein each); 4.5 µl IPG-buffer; 45 µl 20% DTT, dissolved in rehydration buffer (7 M urea, 2 M thiourea, 4% [wt/vol] CHAPS, 1 µl bromophenol blue); the total volume was filled up to 450 µl with rehydration buffer. The resulting protein solution was used for passive rehydration of immobilised pH gradient (IPG) strips with a pH range of 4 to 7 and a length of 24 cm. Isoelectric focusing was performed in an Ettan IPGphor 3 device (GE Healthcare, Munich, Germany) according to the manufacturer's rehydration loading protocol: after active rehydration at 30 V for 10 h, samples were focused at 500 V for 1 h, followed by 1000 V for 1 h, 10.000 V for 3 h and 10.000 V until 42,500 Vh were reached. Focused IPG strips were stored at -80°C.

## 3.1.7. Second dimension

Focused IPG strips were equilibrated immediately before the second dimension SDS-PAGE run. For this purpose, IPG strips were placed in equilibration tubes and incubated with 10 ml equilibration buffer (75 mM Tris-HCl, pH 8.8; 6 M urea; 3.3 M glycerol; 70 mM SDS; 15 mM bromophenol blue) containing 65 mM DTT (EB1) for 15 min with gentle agitation. EB1 was poured off and IPG strips were incubated in 10 ml equilibration buffer containing 135 mM iodoacetamide (EB2) for 15 min. SDS-PAGE was performed in an Ettan-Dalt II apparatus using 1-mm-thick 12.5 % SDS gels with a size of 20 × 26 cm. Run parameter were as follows: 1 W per gel for 45 min, followed by 17 W per gel for 3.5 h at 20°C.

## 3.1.8. Detection of differentially expressed proteins

SDS-PAGE gels were scanned with a Typhoon Trio laser scanner (GE Healthcare, Munich, Germany) at 100 µm resolution using appropriate filters for the excitation/emission wavelengths of Cy2 (488 nm/520 nm), Cy3 (532 nm/580 nm) and Cy5 (633 nm/670 nm) dyes. The corresponding gel images were analysed with the DeCyder software version 6.5 (GE Healthcare, Munich, Germany). Protein spots were detected, quantified and matched automatically as described by the manufacturer. The accuracy of this process was controlled manually. To quantify the protein abundance ratio, the protein spots detected in the proteome of the samples were related to those of the internal standard (Figure 5).

The proteomes of *E. coli* isolated from small intestine and caecum of mice fed the lactose or the casein diet were compared with those obtained from mice fed the starch diet. Differentially expressed proteins with fold-changes of ≥ 2 and P ≤ 0.05 were given. Fold-changes were expressed as average ratios. Statistical analyses were performed with 1-Way AOV between different groups and Student's T-test according to the manufacturer's instruction.

### 3.1.9. Identification of differentially expressed proteins

Preparative gels with 500 µg of bacterial protein were prepared as described above and stained over night with ruthenium II tris (bathophenanthroline disulfonate) [RABILLOUD, 2001]. Proteins of interest were cut out of the two-dimensional gels using an Ettan-Dalt spot picker (GE Healthcare, Uppsala, Sweden). Tryptic digestion was performed in 96-well plates with V-bottom (Greiner bio one, Frickenhausen, Germany) as follows: excised gel pieces were washed twice with 100 µl 50 mM ammonium hydrogen carbonate-50% methanol and incubated for 30 min, dehydrated by addition of 100 µl acetonitrile for 10 min and dried under vacuum using a RCT 90 SpeedVac (Jouan Robotics, St Herblain, France). In-gel digestion was performed over night at 37°C by addition of 50 µl trypsin solution (1 ng/µl trypsin, dissolved in 20 mM ammonium hydrogen carbonate; Promega, Heidelberg, Germany). After digestion, gel pieces were dehydrated by addition of 60 µl 0.1% TFA in

50% acetonitrile for 20 min. Supernatants were transferred to Eppendorf tubes and evaporated using a SpeedVac. Peptides were identified by nano-liquid chromatography-electrospray ionisation-tandem mass spectrometry, electrospray ionisation mass spectrometry (ESI-MS) and tandem mass spectrometry (MS-MS) analysis as described by ALPERT, 2005, ALPERT, 2009 and VOGEL-SCHEEL, 2010.

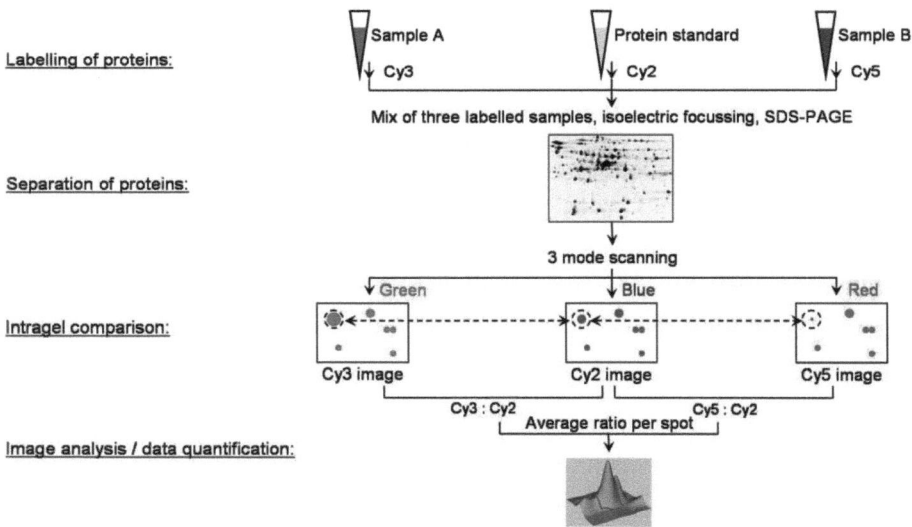

Figure 5.   Principle of the 2D-DIGE analysis. Cy3- and Cy5-labelled samples were mixed with the Cy2-labelled internal standard and separated on a 2D gel. The gels were scanned using appropriate filters for the excitation/emission wavelengths of each dye, which enables separation of the various samples and the internal standard. The protein abundance of the samples was quantified relative to the internal standard. Since the protein concentration in the internal standard is the same on every gel, this method reduces inter-gel variation. Cy, fluorescent cyanine minimal dyes (modified after 2D Electrophoresis Principles and Methods, GE Healthcare, 2004).

### 3.1.10. Determination of dietary substrates in the luminal gut contents

Concentrations of sucrose, fructose, glucose and lactose in intestinal contents of the gnotobiotic mice were determined using enzymatic test kits (Lactose/D-Galactose, Sucrose/D-Glucose/D-Fructose; R-Biopharm,

Darmstadt, Germany). The protocol of the manufacturer was adapted to a 96-well plate format, allowing a 20-fold reduction of the volumes of all test components. Absorption at 340 nm was measured using a Tecan Infinite F200 Pro microplate reader (Tecan Group Ltd., Männedorf, Switzerland). Extinction differences were calculated as follows: $\Delta E = (E2-E1)_{sample} - (E2-E1)_{blank}$. Concentrations were calculated using standard curves.

Protein concentration in the gut contents was determined by the method of Bradford as described above (see section 3.1.4.). Quantification of amino acids was based on the ninhydrin (1,2,3-Indantrione monohydrate, Fluka, Neu-Ulm, Germany) reaction [HWANG, 1975] and was performed as follows using 96-well plates: 60 µl of 20 mM ninhydrin in ethanol and 300 µl luminal gut content were mixed in Eppendorf tubes; background absorption was measured at 570 nm (E1); reaction of ninhydrin with free alpha-amino groups was started by heating at 90°C for 7 min; after cooling on ice, the resulting colour change to deep purple was measured at 570 nm (100 µl each, in duplicates) (E2). Extinction differences were calculated as described above. Concentrations of free alpha-amino groups were calculated using a standard curve (0 to 2.5 g/l hydrolysed casamino acids; Bacto laboratories, Mt Pritchard, NSW, Australia).

Intestinal galacturonate and glucuronate concentrations were measured enzymatically using uronate dehydrogenase (Udh) from *Agrobacterium tumefaciens* [MOON, 2009]. The plasmid pETATu containing Udh was kindly provided by K. L. Jones Prather (MIT, Cambridge, USA). Expression of Udh in *E. coli* BL21 and protein purification was performed as described by MOON *et al.* Small intestinal, caecal and colonic contents (200 µl each) were lyophilised (Martin Christ Gefriertrocknungsanlagen GmbH, Osterode, Germany) and resuspended in 20 µl ddH$_2$O. The enzymatic assays were done in a 96-well plate format as follows: 195 µl Tris-HCl (105 mM, pH 8.0),

4 µl NAD$^+$ (32 mM) and 4 µl sample were incubated at room temperature for 5 min; background absorption was measured at 340 nm (E1); the reaction was started by addition of 1 µl Udh and the mixture was incubated at room temperature for 30 min; the absorbance increase was determined at 340 nm (E2). Extinction differences were calculated as described above. Concentrations of galacturonate and glucuronate were calculated with the help of standard curves (0 to 15 mM D-galacturonic acid sodium salt and D-glucuronic acid sodium salt monohydrate, respectively, Sigma Aldrich, Munich, Germany).

## 3.2. *In vitro* analysis

### 3.2.1. Generation of luciferase reporter gene constructs

To investigate the expression profile of genes of interest (*ahpCF*, *dps*, *kdul*, *kduD*, *uxaCA*, *uxaB* and *uxuAB*) in response to different stimuli, their promoter regions were cloned in front of the bacterial luciferase genes *luxAB* (Figure 6). For this purpose, the promoter regions were amplified from *E. coli* MG1655 by colony PCR using primers flanked by the restriction enzyme sequences of XbaI or EcoRI. To improve the restriction digestion, two additional nucleotides were added at the 5' ends of the primers. The primers and specific annealing temperatures used are listed in Table I, APPENDIX II. Colony PCR was performed in a 48-µl reaction mixture including 2.4 µl of each primer (10 µM, Eurofins MWG Operon, Ebersberg, Germany), 0.76 µl 12.5 mM dNTP (Invitek, Berlin, Germany), 4.8 µl 10 × Dream Taq Green buffer, 1.25 U Dream Taq DNA polymerase (Fermentas, St. Leon-Rot, Germany) and 2 µl PCR template (a single bacterial colony was resuspended in 50 µl ddH$_2$O). PCR conditions were as described above (see section 3.1.3.). The size of the PCR products was controlled in 2.5% agarose gels with the help of the GeneRuler™ 100 bp Plus DNA Ladder (Fermentas, St. Leon-Rot, Germany). PCR products were purified using the High Pure PCR

Product Purification Kit as described by the manufacturer (Roche Applied Science, Mannheim, Germany).

Amplified PCR fragments and the *luxAB*-containing plasmid pKEST-MR (Figure I, APPENDIX II) were digested with XbaI and EcoRI (FastDigest®; Fermentas, St. Leon-Rot, Germany) according to the manufacturer's instructions. Digested plasmids were separated in 1% agarose gels, cut out and purified as described above (see section 3.1.3.), whereas digested PCR fragments were purified using the High Pure PCR Product Purification Kit (Roche Applied Science, Mannheim, Germany). Ligation was done at a vector:insert ratio of 1:5 using the T4 DNA Ligase (New England Biolabs, Beverley, USA). The required amount of DNA was calculated as follows: $mass_{insert}[ng] = 5 \times mass_{vector}[ng] \times size_{insert}[in\ bp] / size_{vector}[in\ bp]$.

The DNA resulting from the ligation reactions was subsequently transformed into *E. coli* MG1655 by electroporation [DOWER, 1988]. Briefly, 2 to 5 µl of the ligation reaction were added to 50 µl of competent cells ($10^8$ bacterial cells/µl), transferred to electroporation cuvettes, and electroporated using a Gene Pulser apparatus (200 Ω, 25 µF, 1.8 kV; BioRad, Munich, Germany). Finally, 950 µl SOC medium (2.5% yeast extract, 10 mM NaCl, 2.5 mM KCl, 10 mM $MgCl_2$, 10 mM $MgSO_4$) was added and the resulting mixture was incubated for 1 h at 37°C. Transformed cells (100 µl each) were plated on LB-agar plates containing carbenicillin (50 µg/ml; Roth, Karlsruhe, Germany) and incubated overnight at 37°C. Transformed colonies were tested for the correct insert size by colony PCR using vector specific primers (5'-AAA GTG CCA CCT GAC GT -3' [sense], 5'-GGG TTG GTA TGT AAG CAA -3' [antisense]) and amplicons were commercially sequenced using the cycle sequencing technology (Eurofins MWG Operon, Ebersberg, Germany).

# MATERIAL AND METHODS

**Figure 6. Schematic presentation of luciferase reporter gene constructs.** The promoter region of genes of interest was cloned in front of the ribosomal binding site (RBS) of the bacterial luciferase genes *luxAB*. Transcription factors specific for the genes of interest led to transcription of the bacterial *luxAB* genes, whose activity was determined by measuring luminescence.

## 3.2.2. Luciferase reporter gene assays

### 3.2.2.1. *ahpCF* and *dps* promoters

To investigate the *ahpCF* and *dps* promoter activity in response to different types of carbohydrates, *E. coli* MG1655 or Δ*oxyR* clones carrying p*ahpCFp::luxAB* or p*dpsp::luxAB* were grown aerobically or anaerobically in LB-Lennox medium plus carbenicillin (50 µg/ml) and inoculated at 5% into 300 ml of fresh LB-Lennox medium plus carbenicillin (50 µg/ml). Fresh cultures were incubated under aerobic and anaerobic conditions for approximately 1.5 h until mid-exponential phase and harvested by centrifugation at 5,000 × g and 4°C for 5 min. Pelleted cells were resuspended in 30 ml LB-Lennox medium plus carbenicillin (50 µg/ml). Cell densities were determined at 600 nm ($OD_{600}$) using a SmartSpec Plus Spectrophotometer (BioRad, Munich, Germany) and cell concentrations were adjusted to $5 \times 10^9$ cells/ml.

To analyse the influence of a variety of test substances on the *ahpCF* and *dps* promoter activities, $H_2O$, $H_2O_2$ (final concentration of 300 µM), various types of carbohydrates (final concentrations of 50 to 400 mM) and NaCl (final concentrations of 400 mM, 700 mM) were applied to sterile 6-well plates for analysis under aerobic conditions or to sterile, gassed Hungate tubes (80% nitrogen, 20% $CO_2$) for analysis under anaerobic conditions. Cell suspensions (1.5 ml each) were added to each well or Hungate tube. To measure the *ahpCF* and *dps* promoter activities in the presence of a protein-rich medium,

1.5 ml cell suspensions were centrifuged at 5,000 × g and 4°C for 5 min, bacterial pellets were resuspended in 1.5 ml SOC medium and samples were added to sterile 6-well plates or to sterile, gassed Hungate tubes (80% nitrogen, 20% $CO_2$).

All samples were shaken at 120 rpm during incubation at 37°C for 30 min. Cells were harvested by centrifugation at 10,000 × g and 4°C for 10 min and bacterial pellets were resuspended in 1.5 ml PBS containing chloramphenicol (30 µg/ml; Roth, Karlsruhe, Germany) to inhibit protein biosynthesis. Luminescence of $2.5 \times 10^8$ cells in 50 µl PBS was measured using a Luminoscan Ascent Luminometer (Labsystems, Helsinki, Finland) in 96-well plates (LuminNunc F96 MicroWell Plate, VWR, Darmstadt, Germany) as follows: 100 µl 2% Decanal (Sigma-Aldrich, Steinheim, Germany) in PBS–10% ethanol was injected automatically to each well, incubated for 3 s and luminescence was measured for 10 s. Each sample was measured in triplicates. Relative luminescence was calculated as follow: absolute luminescence values of stimulated cells were divided by values observed for cells grown on LB-Lennox medium without stimuli.

### 3.2.2.2. *kduI* and *kduD* promoters

To identify substances inducing *kduI* and *kduD* gene expression, *E. coli* MG1655 or Δ*oxyR* clones carrying p*kduIp::luxAB* or p*kduDp::luxAB* were incubated in 10 ml M9 minimal medium ($Na_2HPO_4$, 6 g/litre; $KH_2PO_4$, 3 g/litre; NaCl, 0.5 g/litre; $NH_4Cl$, 1 g/litre; uracil, 12.5 mg/litre; 1 mM $MgSO_4$, 0.1 mM $CaCl_2$) plus carbenicillin (50 µg/ml) in the presence of the pulverised mouse diets (1%), scratched small intestinal mucosal tissue (1%) or the dietary components glucose, fructose, galactose (50 mM each), lactose (25 mM) and casamino acids (1%). Cultures were grown aerobically in 100 ml Erlenmeyer flasks or anaerobically in gassed Hungate tubes (80% nitrogen, 20% $CO_2$) under shaking at 220 and 120 rpm, respectively, and 37°C for 16 h. Cells were harvested, resuspended in PBS containing chloramphenicol (30 µg/ml) and luminescence of $2.5 \times 10^8$ cells was measured as described above. Relative luminescence values were calculated as follows: absolute luminescence values were related to values measured in the presence of glucose, which directly undergoes conversion by glycolytic enzymes and therefore did not induce the *kduI* and *kduD* gene expression.

### 3.2.2.3. *uxaC*, *uxaB* and *uxuAB* promoters

To investigate the expression profile of *uxaCA*, *uxaB* and *uxuAB* in response to osmotic stress, *E. coli* MG1655 or Δ*oxyR* clones carrying p*uxaCAp::luxAB*, p*uxaBp::luxAB* or p*uxuABp::luxAB* were precultured, grown until mid-exponential phase and harvested as described for clones carrying p*ahpCFp::luxAB* and p*dpsp::*luxAB. Pelleted cells were washed twice with 20 ml PBS and resuspended in M9 medium. Cell densities were determined and cell concentrations were adjusted as described above. Cell suspensions (450 µl) were transferred to Eppendorf tubes, centrifuged at 5,000 × g and 4°C for 5 min and bacterial pellets were resuspended in M9 medium (negative control) or M9 medium containing galacturonate or glucuronate (50 mM each) with or without sucrose (25 to 400 mM) or $H_2O_2$ (300 µM) plus carbenicillin (50 µg/ml). Cell suspensions were incubated at 37°C for 90 min under shaking at 120 rpm in sterile 12-well plates for analysis under aerobic conditions or in sterile, gassed Hungate tubes (80% nitrogen, 20% $CO_2$) for analysis under anaerobic conditions. Samples were harvested, resuspended in PBS containing chloramphenicol (30 µg/ml) and luminescence of $2.5 \times 10^8$ cells was measured as described above. Relative luminescence values were calculated as follows: luminescence values were related to values measured in the presence of galacturonate and glucuronate, respectively.

## 3.2.3. Generation of knock-out mutants

Knock-out mutants were generated according to the method of DATSENKO and WANNER [DATSENKO, 2000]. For this purpose, the chromosomal sequences internal to the *ahpCF*, *oxyR* and *kduID* genes were replaced by a kanamycin resistance cassette (Figure 7). Briefly, the kanamycin resistance cassette was amplified from pKD13 using primers containing a pKD13 priming site at the 3' end and a homologous sequence to the target gene at the 5' end (Table I, APPENDIX II). Amplicons were separated in 1% agarose gels, cut out and purified as described above. The purified DNA was introduced into *E. coli* MG1655 competent cells expressing the lambda Red recombinase (encoded on pKD46) [DATSENKO, 2000], which mediates homologous recombination. Cells with a fixed chromosomal mutation were selected on LB-Lennox agar containing kanamycin (50 µg/ml; Roth, Karlsruhe, Germany) and validated by colony PCR with the primers listed in Table I, APPENDIX II. To confirm the genotype, amplicons were commercially sequenced using the cycle sequencing technology (Eurofins MWG Operon, Ebersberg, Germany).

## 3.2.4. Complementation of knock-out mutants

To complement the growth defects of the generated knock-out mutants, the *ahpCF*, *oxyR* and *kduID* genes including their corresponding promoters, were amplified from *E. coli* MG1655 by colony PCR using primers flanked by the restriction enzyme sequences HindIII or BamHI (Table I, APPENDIX II). Colony PCR was performed in a 48 µl reaction mixture as described above (see section 3.2.1.). Amplicons were controlled for their correct size in 1% agarose gels with the help of the 1 Kb Plus DNA Ladder (Invitrogen, Carlsbad, USA) and purified (High Pure PCR Product Purification Kit; Roche Applied Science,

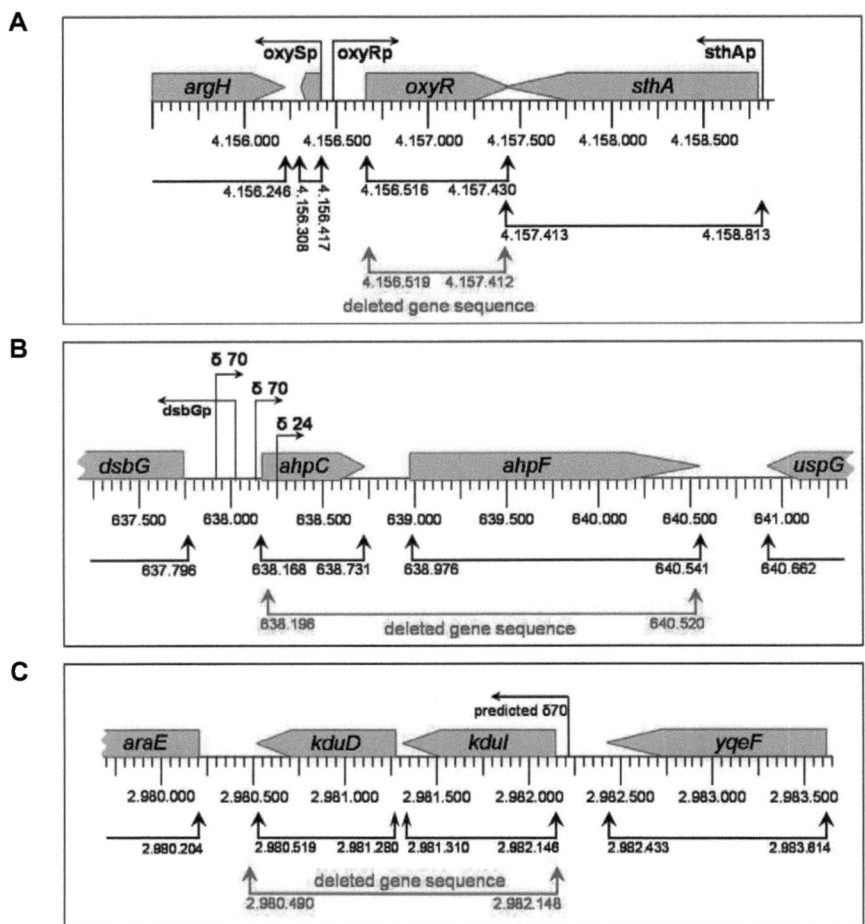

**Figure 7. Deleted chromosomal regions in mutant *E. coli* strains. (A)** *E. coli* ΔoxyR, **(B)** *E. coli* ΔahpCF, **(C)** *E. coli* ΔkduID. The expression and function of the adjacent genes was not affected because promoter and coding regions remained unchanged in the mutants. Black arrows indicate start and termination sites of genes, red arrows indicate deleted sequences.

Mannheim, Germany). Amplified fragments and the low copy vector pSU19 were digested with HindIII and BamHI (FastDigest®; Fermentas, St. Leon-Rot, Germany) according to the manufacturer's instruction. Digested pSU19 plasmids were separated in 1% agarose gels, cut out and purified as described above (see section 3.1.3.), whereas digested PCR fragments were purified with the High Pure PCR Product Purification Kit (Roche Applied Science, Mannheim, Germany). Ligation and transformation in *E. coli* Δ*ahpCF*, Δ*oxyR* or Δ*kduID* competent cells were done as described for the luciferase constructs (see section 3.2.1.). Positive clones (*E. coli* Δ*ahpCF* pSU19*ahpCF*, *E. coli* Δ*oxyR* pSU19*oxyR*, *E. coli* Δ*kduID* pSU19*kduID*) were selected on LB-Lennox agar containing chloramphenicol (10 µg/ml) and tested for the correct insert size by colony PCR using vector specific primers (5'-CCA GGC TTT ACA CTT TAT GC -3' [sense], 5'-AGG CTG CGC AAC TGT TG -3' [antisense]). Amplicons were commercially sequenced using the cycle sequencing technology (Eurofins MWG Operon, Ebersberg, Germany).

### 3.2.5. Growth experiments
### 3.2.5.1. Growth of *E. coli* strains Δ*ahpCF* and Δ*oxyR*

Growth of *E. coli* MG1655 and mutants lacking the *ahpCF* or *oxyR* genes was monitored in the presence or absence of osmotic stress. For this purpose, cells were precultured aerobically or anaerobically in LB-Lennox medium, mutants in the presence of kanamycin (50 µg/ml), and inoculated at $2.5 \times 10^7$ cells/ml into fresh LB-Lennox medium or LB-Lennox medium containing sucrose (400 mM, 700 mM). Incubation at 37°C was done in 100 ml Erlenmeyer flasks (20 ml medium) for aerobic conditions or in gassed Hungate tubes (5 ml medium; 80% nitrogen, 20% $CO_2$) for analysis under anaerobic conditions under constant shaking at 180 rpm. $OD_{600}$ was measured hourly over 8 h, after 24 h and, for anaerobically grown cultures, after 28 and 32 h using a SmartSpec Plus spectrophotometer (BioRad, Munich, Germany). The specific growth rate (µ) was determined by plotting the logarithm of the optical density against time (slope in the exponential growth phase corresponds to µ). The doubling time ($t_d$) in min was calculated as follows: $t_d = \ln2 / \mu \times 60$.

## 3.2.5.2. Growth of *E. coli* strain Δ*kduID*

To characterise the growth of mutants lacking the *kduID* genes, *E. coli* MG1655 and *E. coli* Δ*kduID* were precultured aerobically or anaerobically in M9 medium containing either galacturonate or glucuronate (50 mM each). Mutants were cultured in the presence of kanamycin (50 µg/ml). Fresh M9 medium containing the same substrate with or without sucrose (200 mM, 400 mM or 700 mM) was inoculated at $2.5 \times 10^7$ cells/ml.

For aerobic incubation, cultures were incubated in transparent 12-well culture plates (Sigma-Aldrich, Steinheim, Germany), 1 ml per well, in a Tecan Infinite F200 Pro microplate reader (Tecan Group Ltd., Männedorf, Switzerland) under constant shaking at 218 rpm and 37°C for 45 h. $OD_{600}$ was measured in intervals of approximately 15 min in multiple reads per well as recommended by the manufacturer. Because of differences in the path length of microplate readers and standard spectrophotometers, $OD_{600}$ values obtained with the Tecan microplate reader were plotted against the values determined with a SmartSpec Plus spectrophotometer (path length of 1 cm). Transformed $OD_{600}$ values were calculated by the following formula: $OD_{600\ [transformed]} = OD_{600} / 0.1069$.

For growth under anaerobic conditions, fresh media was inoculated in sterile, gassed Hungate tubes (80% nitrogen, 20% $CO_2$) and transferred to 12-well plates (1 ml per well) in an anaerobic chamber (80% nitrogen, 10% $CO_2$, 10% $H_2$; Meintrup dws Laborgeräte, Lähden-Holte, Germany). The plates were covered with an adhesive film to maintain anaerobic conditions, which were controlled by the redox indicator resazurin (1 µg/ml). Sterility was controlled by incubating medium without bacteria. Incubation was done in a Tecan Infinite F200 Pro microplate reader at 37°C for 40 h. Shaking was performed immediately before $OD_{600}$ measurement (as described above) at 450 rpm for 20 sec. Transformation of $OD_{600}$ values and determination of the specific growth rates and doubling times was done as described for aerobic incubation.

### 3.2.5.3. Growth of complemented *E. coli* strains

To characterise the growth of *E. coli* deletion mutants containing the complementing pSU19 plasmids (*E. coli* ΔahpCF pSU19ahpCF, *E. coli* ΔoxyR pSU19oxyR and *E. coli* ΔkduID pSU19kduID), cells were grown in the presence of chloramphenicol (10 µg/ml) under aerobic conditions as described above.

### 3.2.6. Determination of medium osmolality

Osmolality of the various media used for the reporter gene assays and growth experiments was determined by freezing point depression with an Automatic Osmometer as described by the manufacturer (Knauer, Berlin, Germany). Calibration was done against water and a calibration solution of 400 mOsmol/kg (12.687 g NaCl/kg).

### 3.2.7. Determination of intracellular reactive oxygen species

To measure the formation of intracellular reactive oxygen species, *E. coli* cells were stained with non-fluorescent dihydrorhodamin 123, which is oxidised to fluorescent rhodamine 123 in the presence of intracellular $H_2O_2$. For this purpose, *E. coli* cells were precultured aerobically in LB-Lennox medium and inoculated at 5% into 10 ml fresh medium. Bacterial cells were harvested after 45 min of growth at 37°C by centrifugation at 5,000 × g and 4°C for 5 min and washed twice with 10 ml PBS. The cell number was adjusted to $1.25 \times 10^7$ cells per ml in a volume of 2 ml. Dihydrorhodamin 123 (5 mM in DMSO) was added to a final concentration of 20 µM and samples were shaken at 180 rpm and 37°C for 45 min [HENDERSON, 1993; Görlach, 2007]. To remove non-absorbed dihydrorhodamin 123, stained cells were washed with 2 ml PBS and resuspended in 2 ml M9 minimal medium.

Negative controls were incubated with M9 medium only. Test substances ($H_2O_2$, 600 µm; carbohydrates, 50 mM, 400 mM; casamino acids, 2%) were added to 800 µl cells in 12-well cell culture plates (Sigma-Aldrich, Steinheim, Germany) and samples were incubated in the dark by shaking at 150 rpm and 37°C for 60 min. Cells were washed with PBS and resuspended in 800 µl PBS. Fluorescence of 100 µl aliquots was subsequently measured in triplicates using a Synergy Microplate Reader (excitation wavelength: 485/20 nm, emission wavelength: 528/20 nm; BioTek, Bad Friedrichshall, Germany). To calculate the emission per cell, fluorescence was divided by the viable cell counts.

### 3.2.8. Overexpression of KduI and KduD

To gain insight into the role of KduI and KduD in the conversion of galacturonate and glucuronate, *E. coli* strains overexpressing *kduI*, *kduD* or both were generated. For this purpose, chromosomal regions of the corresponding genes were amplified from *E. coli* MG1655 (Figure 8A) by PCR using the primers listed in Table I, APPENDIX II. PCR was performed in a 48-µl reaction mixture as described above (see section 3.2.1.) and size of the amplicons was controlled in 1% agarose gels with the help of the 1 Kb Plus DNA Ladder (Invitrogen, Carlsbad, USA). Amplified PCR fragments were purified (High Pure PCR Product Purification Kit; Roche Applied Science, Mannheim, Germany) and cloned into the pGEM-T Easy vector (Promega Corporation, Madison, USA) using a vector:insert ratio of 1:3 according to the manufacturer's recommendation. The DNA resulting from the ligation reactions was subsequently transformed into *E. coli* JM109 or KRX competent cells (Promega Corporation, Madison, USA) by heat shock at 42°C [FROGER, 2007]. Briefly, 2 µl of the ligation reaction were added to 50 µl competent cells and incubated for 20 min on ice. Cells were heat-shocked for 45 s at 42°C, subsequently returned on ice for 2 min and resuspended in

950 µl SOC medium. Samples were incubated for 1 h at 37°C under constant shaking at 220 rpm. 100-µl aliquots were subsequently plated onto LB-Lennox agar plates containing carbenicillin (50 µg/ml), IPTG (0.5 mM) and X-Gal (40 µg/ml). After incubation overnight at 37°C, transformed white colonies were tested for the correct insert by colony PCR using vector specific T7 (5'-TAA TAC GAC TCA CTA TAGG G -3') and SP6 (5'-GAT TTA GGT GAC ACT ATA G 3') primers and the cycle sequencing technology (Eurofins MWG -

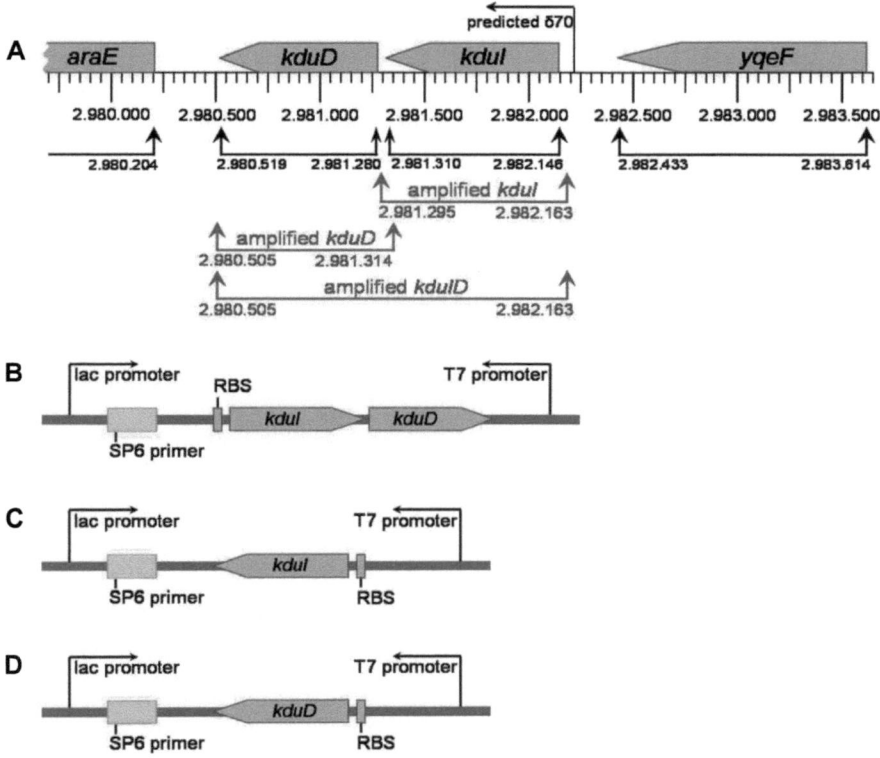

**Figure 8. Schematic presentation of pGEM-T constructs.** The coding regions and ribosomal binding sites (RBS) of the *kduI* and *kduD* genes were amplified from *E. coli* MG1655 and cloned into the multiple cloning site of pGEM-T. Dependent on the orientation of the cloned genes, the expression was under the control of the lacZ- or the T7-promoter. **(A)** Chromosomal regions of the *kduI* and *kduD* genes amplified by PCR. Black arrows, start and termination sites of genes, red arrows, amplified regions. **(B)** pGEM-T-*kduID*. **(C)** pGEM-T-*kduI*. **(D)** pGEM-T-*kduD*.

Operon, Ebersberg, Germany).

To induce protein expression, clones were precultured aerobically in Terrific Broth (TB: tryptone, 12.0 g/litre; yeast extract, 24.0 g/litre; glycerol, 4 ml; 89 mM potassium phosphate, pH 7.5) plus carbenicillin (50 µg/ml) and inoculated at 1% into 10 ml fresh TB plus carbenicillin (50 µg/ml). Cultures were shaken at 220 rpm and 37°C until an $OD_{600}$ of 0.8-1.0 was reached. Protein expression was induced by addition of rhamnose (0.1%) or IPTG (1 mM), dependent on the orientation of the cloned genes (Figure 8B-D), and cultures were incubated at 220 rpm and 25°C for 16 h. Protein expression was controlled by SDS-PAGE [LAEMMLI, 1970] (Figure 9).

### 3.2.9. Determination of hexuronate conversion in the presence of KduI and KduD

To investigate the possible role of KduI and KduD in galacturonate and glucuronate conversion, the corresponding genes were overexpressed as described above. $OD_{600}$ was measured using a SmartSpec Plus spectrophotometer (BioRad, Munich, Germany). Cells were harvested by centrifugation at 5,000 × g and 4°C for 5 min and bacterial pellets were washed twice with 2 ml sodium phosphate buffer (100 mM, pH 7.0) containing a protease inhibitor mix (1:25 dilution of a 25 × stock, Roche Diagnostics GmbH, Mannheim, Germany). Cell concentration was adjusted to approximately $2.5 \times 10^{10}$ cells/ml in a total volume of 2 ml. Cell disruption was done with a FastPrep-24 instrument (MP Biomedicals Germany, Eschwege, Germany) and protein concentration of cell-free extracts was measured with a Bradford assay as described in section 3.1.4.

The conversion of hexuronates was monitored as follows: 5.4 µl of 107 mM galacturonate or 107 mM glucuronate and 2.3 µl of 250 mM NADH (10 mM final concentration each) were added to 50 µl cell-free extract on ice; the reaction was started by incubation at 37°C; samples were taken at 1, 2, 3, 4 and 6 h. The reaction was stopped by addition of trichloroacetic acid (TCA) as follows: 10 µl TCA (100 w/v) were added to 50 µl cell-free extract (17% final concentration) and incubated for 30 min on ice; denaturated proteins were removed by centrifugation at 10,000 × g and 4°C for 10 min and supernatants were collected. Concentrations of galacturonate and glucuronate were determined enzymatically as described above (see section 3.1.10.) and referred to mg protein. Prior to the experiment, it was excluded that the TCA present in the samples interfere with the Udh assay. Specific activities of cell-free extracts were calculated as follows: substrate concentration [nmol] / incubation time [min] / protein concentration [mg/ml].

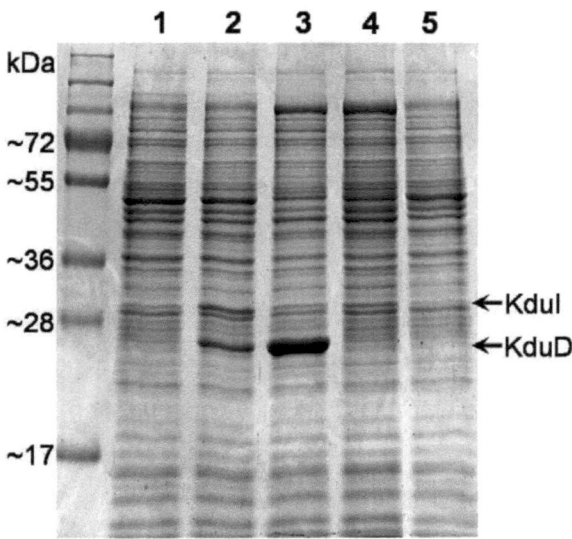

**Figure 9.** **Overexpression of KduI and KduD.** SDS-PAGE of cell-free extracts of *E. coli* carrying the empty pGEM-T or pGEM-T containing *kduI*, *kduD* or both genes. Dependent on the orientation of the cloned genes, the expression was under the control of the lacZ- (JM109 strain) or the T7-promoter (KRX strain). Per lane 10 µg protein was applied. Black arrows indicate the band of the overexpressed proteins. **Lane 1**, *E. coli* JM109 pGEM-T (negative control). **Lane 2**, *E. coli* JM109 pGEM-T-*kduID*. **Lane 3**, *E. coli* KRX pGEM-T-*kduD*. **Lane 4**, *E. coli* KRX pGEM-T-*kduI*. **Lane 5**, *E. coli* KRX pGEM-T (negative control).

## 3.2.10. Determination of intracellular hexuronate concentration

To determine the intracellular hexuronate concentration of *E. coli* during growth on carbohydrates, *E. coli* MG1655 were precultured aerobically in 10 ml M9 minimal medium with glucose (50 mM) and lactose (25 mM), respectively, and inoculated at 2.5% into 20 ml fresh M9 minimal medium containing the same carbon source. Cultures were incubated aerobically at 37°C under shaking at 220 rpm. Samples (2.5 ml) were taken at 2, 3, 4, 6, and at 16 h. Cell density was determined by measuring $OD_{600}$. Cells were harvested by centrifugation at 10,000 × g and 4°C for 5 min, washed and resuspended in 250 µl sodium phosphate buffer (100 mM, pH 7.0). Cell disruption was done in a FastPrep-24 (MP Biomedicals Germany, Eschwege, Germany) instrument and protein concentration was determined by the method of Bradford as described above (see section 3.1.4.). Cellular proteins were removed by TCA precipitation (7% final concentration) as described in see section 3.2.9. Supernatants (95 µl) were transferred to 96-well plates in duplicates and lyophilised (Martin Christ Gefriertrocknungsanlagen GmbH, Osterode am Harz, Germany). Galacturonate and glucuronate concentrations were measured as described in section 3.1.10. Because of the low cell density of anaerobically grown *E. coli*, this experiment could only be performed under aerobic conditions.

## 3.3. Statistical analysis

Statistical analyses were performed with GraphPad Prism 5 (GraphPad, La Jolla, USA). Data were tested for normal distribution with the D'Agostino and Pearson omnibus normality test and the Kolmogorow-Smirnow test. Normally distributed data were given as means ± standard deviation and tested by One-way ANOVA (1-Way AOV) and Dunnett's multiple comparison test for statistically significant differences. Non-normally distributed data were presented as medians and minima versus maxima or interquartile ranges (25% to 75%) and tested by the Kruskal-Wallis 1-Way AOV and the Dunn's multiple-comparison test or the Mann-Whitney-U-test (U-test) for statistically significant differences. Correlation analyses were done with SPSS 16 (SPSS, Inc., Chicago, USA) by using the Spearman's rank correlation coefficient.

# 4. Results

## 4.1. Characterisation of the gnotobiotic mouse model

### 4.1.1. Cell numbers of intestinal *E. coli*

The objective of this work was the identification of mechanisms that enable intestinal bacteria to adapt to nutritional factors in the mammalian intestine. For this purpose, gnotobiotic mice monoassociated with *E. coli* MG1655 were fed three different diets: a diet rich in starch, a diet rich in non-digestible lactose and a casein-rich diet. In order to specify the characteristics of this model, viable cell numbers of intestinal *E. coli* were determined. In all feeding groups, *E. coli* numbers in the small intestine were 12 to 15-fold lower than in the caecum. On the starch and the casein diet, the small intestinal *E. coli* counts were 9 to 19-fold lower than those of the colon, whereas there was no difference on the lactose diet. Mice fed the lactose diet had 8 to 16-fold higher *E. coli* counts in their small intestine than mice fed the starch or the casein diet. Furthermore, caecal cell counts were 14-fold higher on the lactose diet compared to those on the casein diet (Figure 10).

### 4.1.2. 16S rRNA gene sequencing profile

In order to exclude contamination of the gnotobiotic mice, 16S rRNA gene analyses of bacterial samples obtained from caecal contents were performed. Due to limitations in sample material, only a subset of animals was analysed (n = 11). Since all mice were housed in the same isolator, this analysis was representative of the microbial status of all animals in the experiment. The absence of double peaks in the sequencing profiles (Figure II, Figure III APPENDIX II) demonstrate that only one bacterial species was present in the intestines of the gnotobiotic mice.

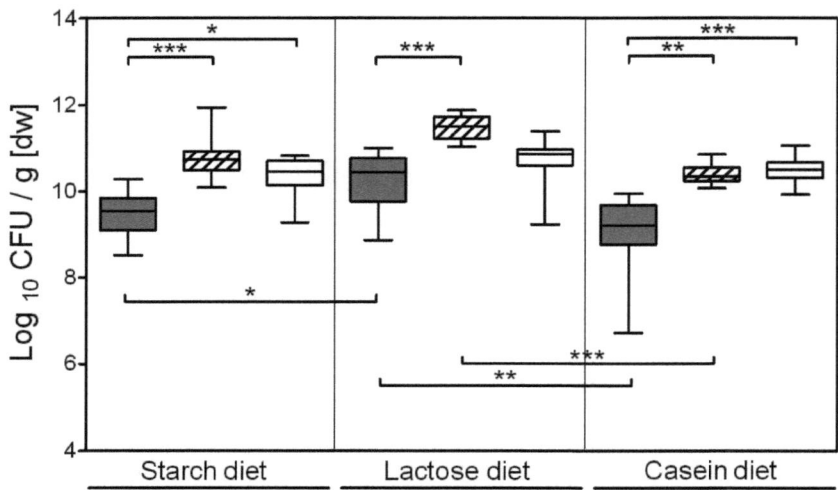

**Figure 10.** **Colony forming units (CFU) of intestinal *E. coli* after 3 weeks of feeding mice diets rich in starch, lactose or casein.** Grey bars, small intestine; hatched bars, caecum; white bars, colon. Data are expressed as medians and minima versus maxima (n = 18 to 21 per diet). Differences between groups were calculated by the Kruskal-Wallis 1-Way AOV and Dunn's multiple-comparison test. *, $P < 0.05$; **, $P < 0.01$; ***, $P < 0.001$.

### 4.1.3. Intestinal substrate availability

To investigate whether feeding of the various diets affects substrate availability in the gut, the intestinal concentrations of sucrose, fructose, glucose, lactose, proteins and amino acids were determined. Since *E. coli* is devoid of enzymes needed for the degradation of polysaccharides or complex proteins [MILLER, 1975; FREUNDLIEB, 1986], in the mouse intestine, only oligo- and monosaccharides or short peptides and amino acids, which result from digestion of complex dietary compounds by host enzymes, are potential carbon and energy sources for this bacterium.

The main source of dietary carbohydrates in this experiment was starch. The highest starch concentrations were present in the starch (43% [wt/wt]) and in the lactose diet (33% [wt/wt]), whereas the casein diet contained only 3% [wt/wt] starch. Starch degradation by host enzymes results in the formation of malto-oligosaccharides and glucose. Accordingly, the highest glucose concentrations were determined in the small intestines of mice fed the starch and the lactose diet (Figure 11A). Moreover, the small intestinal glucose concentration of mice fed the starch diet was three-fold higher than that of mice fed the lactose diet (6 mM vs. 2 mM). In contrast, intestinal contents of mice fed the casein diet contained < 0.5 mM glucose.

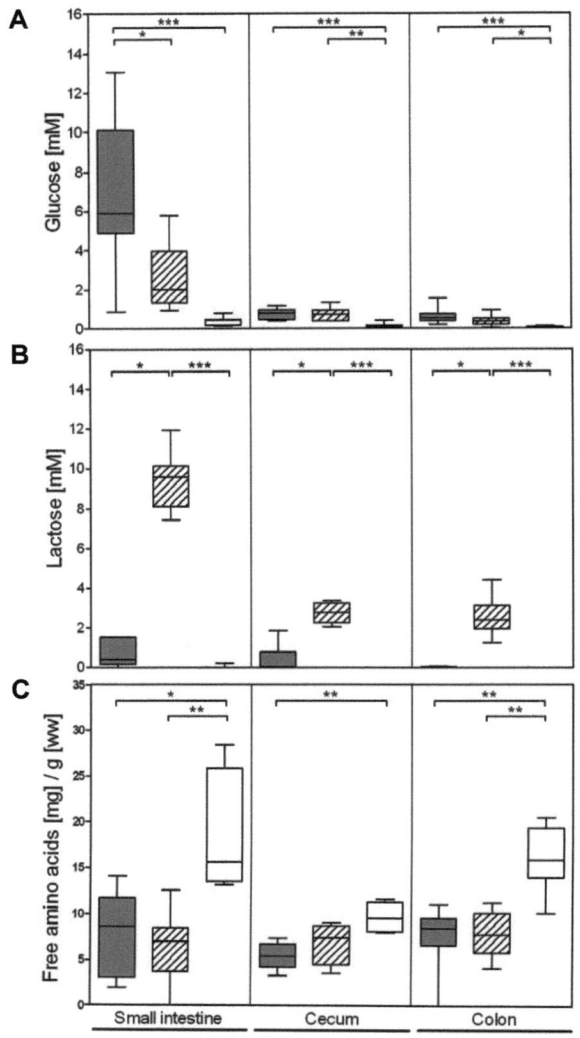

Figure 11. Concentrations of glucose, lactose and amino acids in the gut contents of mice fed the starch, the lactose or the casein diet. Grey bars, starch diet; hatched bars, lactose diet; white bars, casein diet. Data are expressed as medians and minima versus maxima. Differences between groups were calculated by the Kruskal-Wallis 1-Way AOV and the Dunn's multiple-comparison test.
*, $P < 0.05$; **, $P < 0.01$; ***, $P < 0.001$.

(A) Glucose concentrations; starch diet, n = 14; lactose diet, n = 10; casein diet, n = 7.

(B) Lactose concentrations, n = 7.

(C) Free amino acid concentrations, n = 8.

Lactose, which is indigestible by adult rodents [TROELSEN, 1992; KAUR, 2006], was detected only in intestinal contents of mice fed the lactose diet (10% [wt/wt]) (Figure 11B). Sucrose, which is released to glucose and fructose by host enzymes, was present in any of the diets (each 20% [wt/wt]). However, the concentrations of sucrose and fructose in the intestines of mice on any of the diets were below 1 mM (data not shown). Therefore, sucrose and fructose were rapidly absorbed by the host and were not expected to significantly affect the growth of *E. coli* in this *in vivo* experiment.

The casein diet contained the highest concentration of dietary protein (60% [wt/wt]). But mice fed the casein diet had only marginally higher protein concentrations in their intestines (4 to 8 mg per g [ww]) compared to mice fed the starch or the lactose diet (each 20% dietary protein [wt/wt]; 3 to 6 mg of protein per g [ww]). In contrast, the concentration of amino acids in the intestines of mice fed the casein diet was 2-fold higher than in the intestines of mice fed the starch or the lactose diet (Figure 11C). Therefore, the intestinal substrate availability was affected by the diets fed to the mice.

## 4.2. Bacterial adaptation to the host diets

### 4.2.1. Identification of differentially expressed *E. coli* proteins

To identify bacterial proteins that were differentially expressed in response to the various diets, 2D-DIGE, electrospray ionisation mass spectrometry and tandem mass spectrometry were applied (for representative 2D gel image see Figure 12). The proteomes of *E. coli* isolated from small intestine and caecum of mice fed the lactose or the casein diet were compared with those obtained from mice fed the starch diet. In total, over one hundred differentially expressed bacterial proteins were identified ($\geq$ 2- fold, $P < 0.05$; Table II, APPENDIX II). For some of these proteins, different isoforms were detected.

# RESULTS

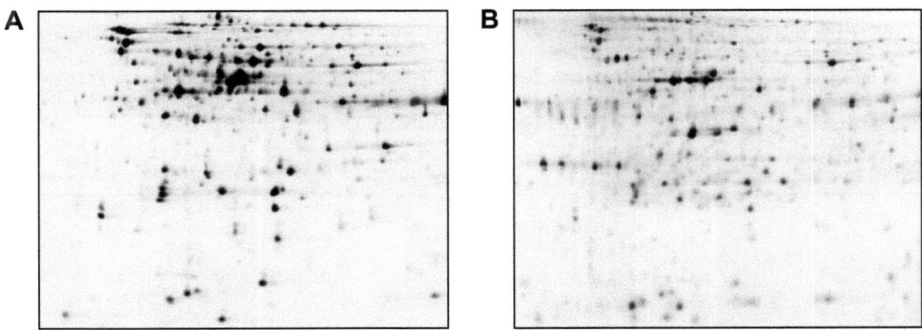

**Figure 12. Representative two-dimensional gel images of the proteome of intestinal *E. coli*** isolated from caecum **(A)** and small intestine **(B)** of mice fed the lactose diet.

First, the differentially expressed proteins identified through this comparison were classified based on their physiological function: some of these proteins, such as the outer membrane proteins A, C and F (OmpA, OmpC, OmpF), are involved in the transport of carbohydrates, ions, peptides and amino acids. However, the majority of the identified proteins are involved in carbohydrate, amino acid and lipid metabolism. For instance, glycolytic enzymes, such as the glyceraldehyde-3-phosphate dehydrogenase (GapA), the pyruvate kinases I (PykF) and II (PykA) and the phosphoglycerate kinase (PgK), as well as enzymes catalysing the breakdown of amino acids, such as the 2-amino-3-ketobutyrate coenzyme A ligase (Kbl) and the serine hydroxymethyltransferase (GlyA), were detected. The biosynthesis of fatty acids and phospholipids was represented by the enoyl-[acyl-carrier-protein] reductase (FabI) and the acetyl-CoA carboxylase (FabE). Enzymes required for transcription, translation and protein folding were also identified, for example elongation factors (EF-Tu, EF-Ts), 30S ribosomal proteins (RpsD, RpSD, RpsJ) and tRNA synthetases (HisS, GltX). Furthermore, enzymes involved in purine and pyrimidine biosynthesis (PurC, PurE, PurH, PyrB) as well as proteins controlling the cell redox homeostasis and stress response were differentially expressed (Figure 13; Table II, APPENDIX II).

This functional classification was used to systematically analyse the roles of the differentially expressed bacterial proteins for the adaptation of E. coli to the intestinal conditions of mice fed the lactose or the casein diet.

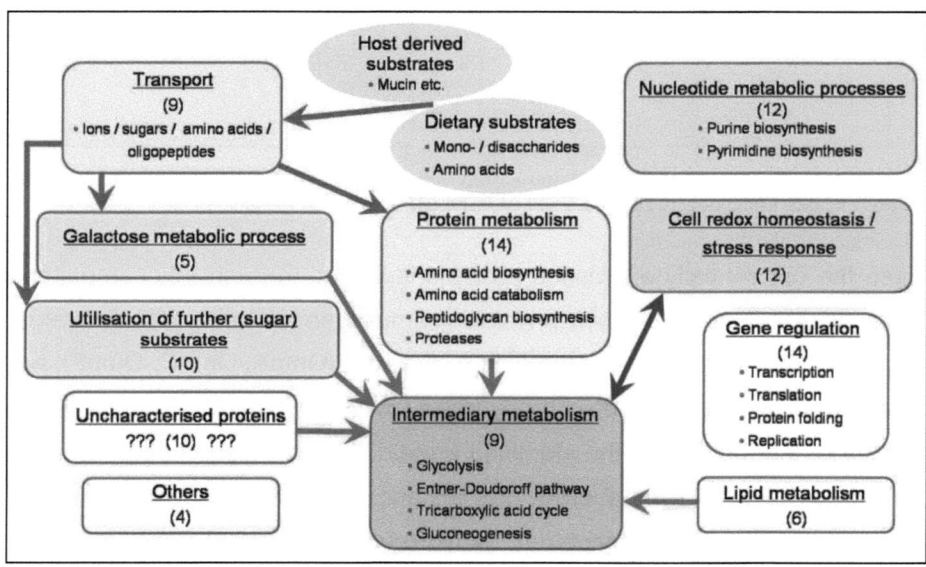

**Figure 13.    Functional categories of differentially expressed *E. coli* proteins.** Proteomes of small intestinal and caecal *E. coli* of mice fed the lactose and the casein diet were compared with those obtained from mice on the starch diet. Numbers of proteins between brackets.

## 4.2.2.    Proof of principle:
## Adaptation of intestinal *E. coli* to the host diets

*E. coli* obtained from mice fed the casein diet repressed enzymes involved in carbohydrate catabolic processes: for example, the glyceraldehyde-3-phosphate dehydrogenase A (GapA), 2.4-fold; the 2-dehydro-3-deoxygluconokinase (KdgK), 2.4-fold; and the pyruvate kinase II (PykA), 2.8-fold. However, enzymes required for peptide and amino acid breakdown were hardly detectable on the casein diet (Figure 13; Table II, APPENDIX II).

In contrast, *E. coli* obtained from mice fed the lactose diet induced enzymes required for amino acid biosynthetic processes: for example, the glutamate dehydrogenase (GdhA), 2.3-fold, and the carbamoyl-phosphate synthase (CarA), 3-fold. Enzymes belonging to the Leloir pathway [HOLDEN, 2003], which is required for the catabolism of galactose, were also upregulated on the lactose diet: the galactose mutarotase (GalM), 4.3-fold; the galactokinase (GalK), 2.3- to 6.6-fold; the galactose-1-phosphate uridylyltransferase (GalT), 2.2- to 7.8-fold; and the UDP-glucose 4-epimerase (GalE), 3.6-fold (Figure 14). The opposite was true for enzymes needed for peptide or amino acid catabolism: the glutaminase 1 (GlsA1) was downregulated 3-fold on the lactose diet in comparison to the starch diet. This reflects the adaptation of the *E. coli* metabolism to the respective host diet.

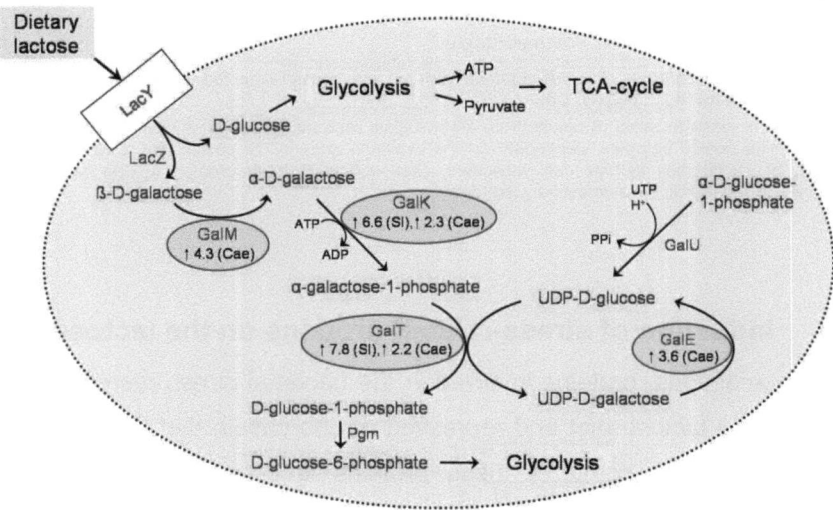

Figure 14. Induction of Leloir pathway enzymes in *E. coli* of mice fed the lactose diet. Proteomes of *E. coli* isolated from small intestine (**SI**) and caecum (**Cae**) of mice fed the lactose diet were compared with those obtained from mice fed the starch diet. Numbers indicate fold-changes. **LacY**, lactose permease; **LacZ**, ß-galactosidase; **GalM**, galactose mutarotase; **GalK**, galactokinase; **GalT**, galactose-1-phosphate uridylyltransferase; **GalE**, UDP-glucose 4-epimerase; **GalU**, uridylyltransferase; **Pgm**, phosphoglucomutase.

RESULTS

Table 2. Regulation of bacterial proteins belonging to cell redox homeostasis and stress-response processes with expression change ≥ 2-fold (P < 0.05) [a].

| Swiss-Prot accession no. | Gene | Protein description | Fold change [b] | | | |
|---|---|---|---|---|---|---|
| | | | Lactose diet versus Starch diet | | Casein diet versus Starch diet | |
| | | | SI | Cae | SI | Cae |
| P0A9A9 | fur | Ferric uptake regulatory protein | | 3.1 | | |
| P0ABT2 | dps | DNA protection during starvation protein | | 3.2 | | |
| P68066 | grcA | Autonomous glycyl radical cofactor | 2.3 | - 8.8 | | |
| P0AFF6 | nusA | Transcription elongation protein nusA | 3.1 | | | |
| P0A6H5 | hslU | ATP-dependent protease ATPase subunit HslU | | | - 4.5 | |
| P05055 | pnp | Polyribonucleotide nucleotidyltransferase | | - 6.6 | | |
| P0A9D2 | gst | Glutathione S-transferase | | | | - 2.7 |
| P35340 | ahpF | Alkyl hydroperoxide reductase subunit F | 3.2 | 2.2 | - 3.5 | |
| P0AE08 | ahpC | Alkyl hydroperoxide reductase subunit C | | | - 2.4 | - 2.1 |
| P0A862 | tpx | Thiol peroxidase | | | | 2.0 |
| P0ACE0 | hybC | Hydrogenase-2 large chain | 2.6 | 2.0 | | |
| P38489 | nfnB | Oxygen-insensitive NAD(P)H nitroreductase | 2.2 | | | |
| P39315 | qorB | Quinone oxidoreductase 2 | | - 4.8 | | |

[a] Comparison of the proteomes of E. coli obtained from the intestines of mice fed the lactose or the casein diet with those of mice fed the starch diet.
[b] Data represent average ratios of results from 20 biological replicates per diet. 2D-DIGE analyses were performed using pooled samples, dependent on the available material. SI casein diet, Cae starch diet: duplicates, SI starch diet, lactose diet: triplicates, Cae lactose diet: quadruplicate, Cae casein diet: quintuplicate, P ≤ 0.05. SI, small intestine; Cae, caecum.

### 4.2.3. Induction of stress-related proteins on the lactose diet

With few exceptions, proteins involved in the bacterial stress response were induced on the lactose diet and repressed on the casein diet (Table 2). The gene expression of some of these proteins is controlled by the OxyR transcriptional dual regulator (OxyR), a known sensor for oxidative stress [Storz, 1990b]. Identified proteins belonging to these regulon include the ferric uptake regulatory protein (Fur), the alkyl hydroperoxide reductase (AhpR), subunits F (AhpF) and C (AhpC) and the DNA protection during starvation protein (Dps). Fur, AhpF and Dps were induced in the intestines of mice fed the lactose diet by factors of 2.2 to 3.2, while AhpC and AhpF were repressed by factors of 2.1 to 3.5 on the casein diet (both in comparison to the starch diet).

### 4.2.4. Regulation of the uncharacterised *E. coli* proteins KduI and KduD

Two proteins, which have been identified in *E. coli* [DUNTEN, 1998; CROWTHER, 2005; HU, 2010) but whose functions are still obscure, were also differentially expressed: the 2-deoxy-D-gluconate 3-dehydrogenase (KduD) was 2.4-fold induced on the lactose diet and 4.0-fold repressed on the casein diet, while the 5-keto 4-deoxyuronate isomerase (KduI) was 8.3-fold repressed on the casein diet (Figure 15). These enzymes are involved in pectin degradation in the plant pathogen *Erwinia chrysanthemi* [CONDEMINE, 1986; CONDEMINE, 1991), but *E. coli* K-12 lacks enzymes required for transport and catabolism of pectin, poly- and oligogalacturonates [RODIONOV, 2004] and is therefore not capable of degrading these carbohydrates. For that reason, KduI and KduD may have different functions in *E. coli*.

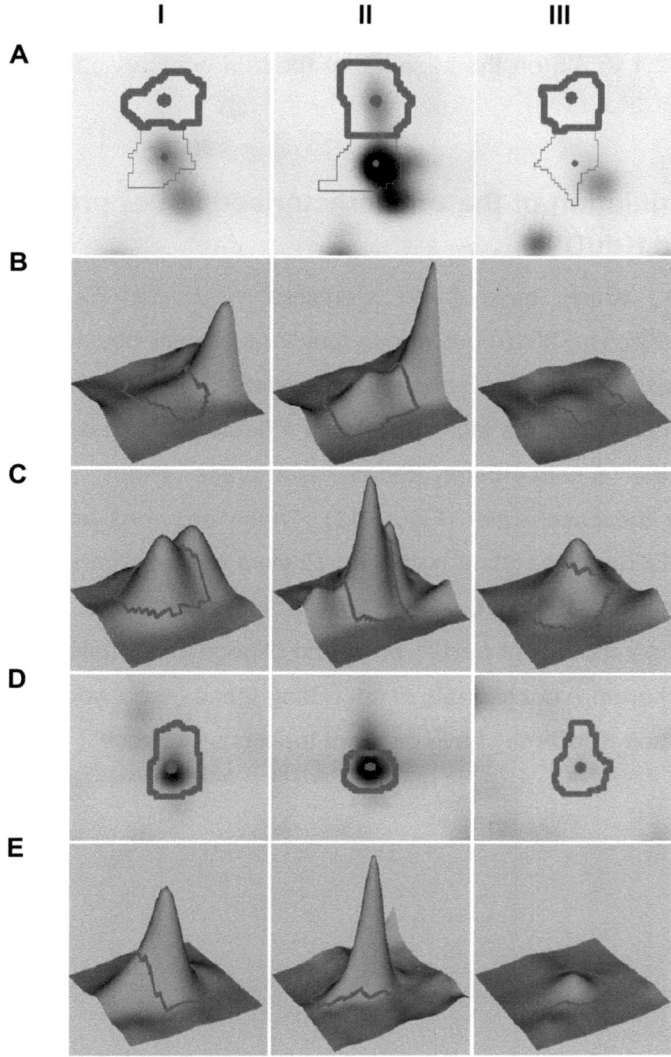

**Figure 15. Regulation of KduD and KduI in intestinal *E. coli*.** Two-dimensional gel images and three-dimensional representation of spots identified as 2-deoxy-D-gluconate 3-dehydrogenase (KduD) and 5-keto 4-deoxyuronate isomerase (KduI) are shown. Representative spots obtained from caecal *E. coli* proteomes of mice fed the starch (I), the lactose (II), or the casein diet (III). **(A)** Two-dimensional gel images of the two identified isoforms of KduD. Thick line, KduD1; thin line, KduD2. **(B, C)** The three-dimensional representation of KduD1 **(B)** and KduD2 **(C)** indicates the induction of this protein on the lactose diet and its repression on the casein diet (both in comparison to the starch diet). **(D)** Two-dimensional gel images of the spot identified as KduI. **(E)** The three-dimensional representation of KduI demonstrates the repression of this protein on the casein diet in comparison to the starch diet.

RESULTS

## 4.3. *In vitro* characterisation of selected proteins

In the following experiments, selected differentially expressed bacterial proteins were characterised *in vitro* for their possible role in the bacterial adaptation to the various host diets.

### 4.3.1. Induction of OxyR-dependent proteins by a lactose-rich diet

#### 4.3.1.1. Induction of the oxyR regulon by osmolytes

The induction of oxidative stress-response proteins belonging to the *oxyR* regulon (AhpF, Dps and Fur) on the lactose diet was an unexpected finding. The expression of the corresponding genes is controlled by the transcriptional regulator OxyR, which is activated by $H_2O_2$, a known inducer of toxic hydroxyl radicals by Fenton or Haber–Weiss reactions [Storz, 1990a;b; Storz, 1990c]. The oxygen partial pressure, which may provoke the formation of reactive oxygen species such as $H_2O_2$ in the mouse intestine, was not assumed to differ in response to substrate availability, the only parameter changed between the three mouse groups. Therefore, other environmental factors than oxidative stress must have triggered the expression of these stress-related proteins on that diet.

The analysis of the intestinal substrate availability (see section 4.1.3.) revealed two major differences in the composition of the intestinal contents of lactose- and casein-fed mice: in the intestines of mice on the casein diet, high amino acid concentrations but no carbohydrates were present, whereas mice on the lactose diet had low amino acid concentrations and high carbohydrate concentrations, especially lactose, in their intestines. Therefore, carbohydrates were assumed to induce the gene expression of proteins belonging to the *oxyR* regulon. To test this hypothesis, the influence of various carbohydrates on the *ahpCF* and *dps* promoter activity was investigated using luciferase reporter gene assays.

These analyses revealed that increasing concentrations of glucose, lactose, sucrose and sorbitol (50 to 400 mM) induced the *ahpCF* and *dps* promoters under both aerobic and anaerobic growth conditions (Figure 16). Activation of the *dps* promoter was higher under anaerobic than aerobic growth conditions (aerobic, up to 7-fold; anaerobic, up to 12-fold), whereas the *ahpCF* promoter was induced equally under aerobic and anaerobic growth conditions (up to

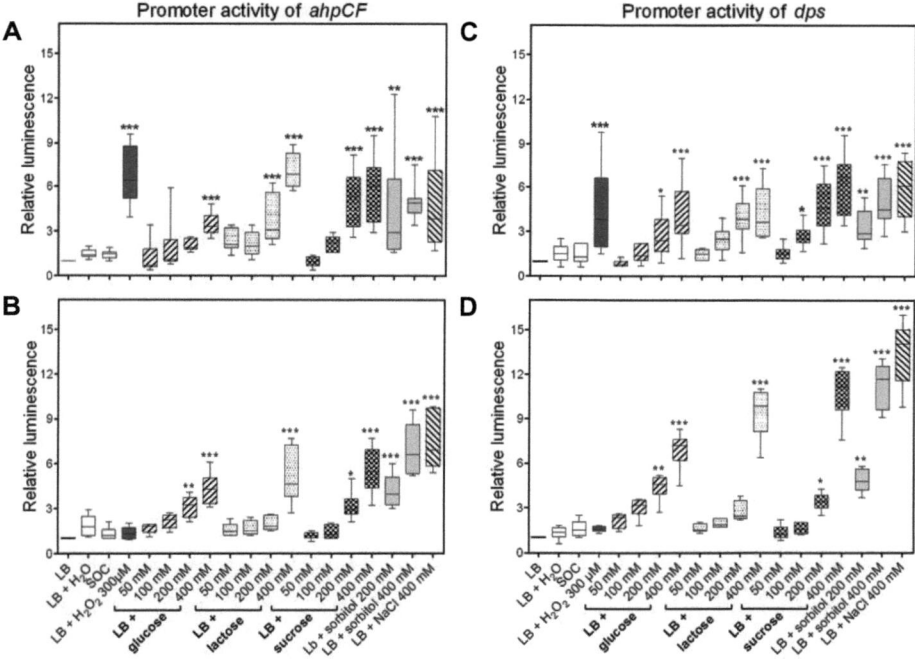

**Figure 16. Induction of the *ahpCF* and *dps* promoters by osmolytes.** Relative luminescence of *E. coli* carrying either p*ahpCFp::luxAB* (A, B) or p*dpsp::luxAB* (C, D) in response to various concentrations of carbohydrates and NaCl after 30 min of incubation under aerobic (**A, C**) and anaerobic (**B, D**) conditions at 37°C are presented. Luciferase activity was normalised to values determined for cells grown on LB medium without stimuli (LB). Data are expressed as medians and minima versus maxima (aerobic: n = 11, anaerobic: n = 6). Differences between treatment groups were calculated by the Kruskal-Wallis 1-Way AOV and the Dunn's multiple-comparison test. *, $P < 0.05$; **, $P < 0.01$; ***, $P < 0.001$.

7-fold). $H_2O_2$, the known activator of OxyR [STORZ, 1990a], was used as a positive control. Under aerobic conditions, incubation of clones with 300 µM $H_2O_2$ increased the activity of both promoters up to 7-fold. $H_2O_2$ did not change the promoter activities in an anaerobic environment, because the added $H_2O_2$ was destabilised by the reducing conditions. Since WEBER et al. demonstrated increased levels of AhpC and Dps in response to NaCl stress (700 mM) [WEBER, 2006], NaCl was also used as a positive control. Stimulation with 400 mM NaCl resulted in similar ahpCF and dps promoter

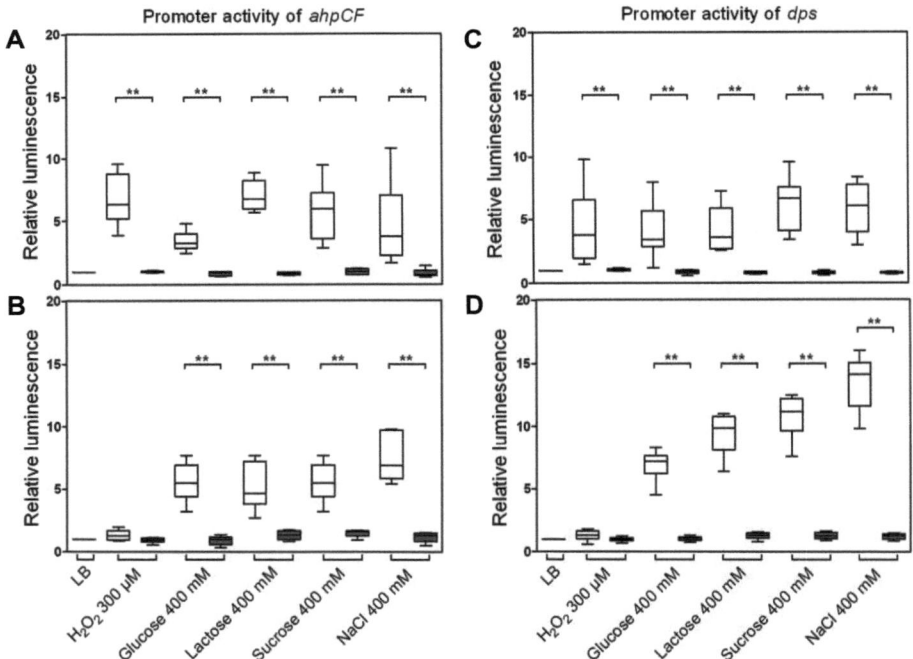

Figure 17. OxyR-dependency of osmolyte-induced *ahpCF* and *dps* expression. Relative luminescence of wild type and *oxyR* deficient *E. coli* carrying either p*ahpCFp::luxAB* (A, B) or p*dpsp::luxAB* (C, D) in response to various carbohydrates and NaCl after 30 min of incubation under aerobic (A, C) and anaerobic (B, D) conditions at 37°C are compared. White bars, wild type *E. coli*; grey bars, *E. coli* ΔoxyR. Luciferase activity was normalised to values determined for cells grown on LB medium without stimuli (LB). Data are expressed as medians and minima versus maxima (n = 6). Differences between wild type and mutant *E. coli* were calculated by the U- test. *, $P < 0.05$; **, $P < 0.01$; ***, $P < 0.001$.

activities as observed in response to $H_2O_2$ and high concentrations of carbohydrates (aerobic, up to 6-fold; anaerobic, up to 14-fold). However, addition of 700 mM NaCl did not further increase the activity of both promoters (data not shown). The transfer of the *E. coli* cells to the protein-rich SOC medium, which mimics the intestinal conditions of casein-fed mice, as well as the addition of water did not change the promoter activity under aerobic or anaerobic conditions and therefore served as negative controls. These results demonstrate that not only $H_2O_2$ but also osmolytes, such as carbohydrates and NaCl, act as inducers of the *ahpCF* and *dps* gene expression.

To clarify whether or not the *ahpCF* and *dps* induction by osmolytes is regulated by OxyR, as described for $H_2O_2$, luciferase reporter gene assays were performed in *oxyR* deficient *E. coli*. In the absence of OxyR, addition of glucose, lactose, sucrose or NaCl (400 mM each) did not change the *ahpCF* and *dps* promoter activity compared to incubation of clones on LB medium without addition of stimuli under both aerobic and anaerobic conditions (Figure 17). This strongly indicates that the expression of *ahpCF* and *dps* by osmolytes was directly mediated by the transcriptional regulator OxyR.

## 4.3.1.2. Positive correlation of *ahpCF* and *dps* expression and medium osmolality

The primary feature of carbohydrates or NaCl is their effect on the medium osmolality. Depending on the concentration, addition of these solutes to LB medium strongly increased the osmolality (LB medium, 230 mOsmol/kg; LB medium plus carbohydrates: 100 mM, ~340 mOsmol/kg; 200 mM, ~430 mOsmol/kg; 400 mM, ~600 mOsmol/kg). Because of the dissociation into its constituent ions, NaCl increased the medium osmolality more strongly than equal concentrations of carbohydrates (LB medium plus NaCl: 400 mM, 1000 mOsmol/kg; 700 mM, 1450 mOsmol/kg; Figure VI, APPENDIX II).

RESULTS

To investigate a possible correlation between the *ahpCF* and *dps* promoter activity and the osmolality of carbohydrates and NaCl, luciferase values were plotted against the observed medium osmolality. The activity of both promoters was positively correlated with the medium osmolality under aerobic as well as under anaerobic conditions (Figure 18).

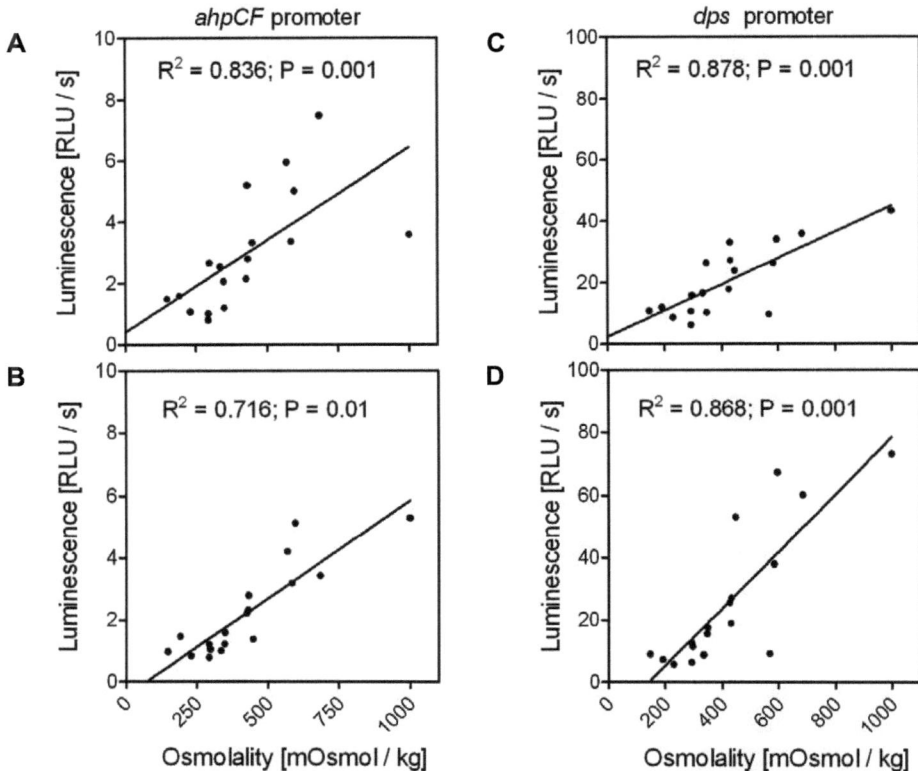

**Figure 18. Positive correlation of *ahpCF* and *dps* promoter activity and medium osmolality.** Data are expressed as medians for promoter activity observed in *E. coli* and means for the osmolality of the various media. The Spearman's rank correlation coefficient was used for statistical analysis. **(A)** For *ahpCF* promoter, aerobic growth conditions; **(B)** for *ahpCF* promoter, anaerobic growth conditions; **(C)** for *dps* promoter, aerobic growth conditions; **(D)** for *dps* promoter, anaerobic growth conditions.

## 4.3.1.3. No formation of intracellular $H_2O_2$ by osmotic stress

Various stress conditions caused by heat, osmolality or chemicals such as ethanol are proposed to result in the formation of intracellular reactive oxygen species, such as $H_2O_2$, $OH^{\bullet}$ and $O_2^-$ ("suicide through stress" theory by ALDSWORTH and DODD [ALDSWORTH, 1999; DODD, 2007]). Since the OxyR transcriptional regulator is activated by $H_2O_2$ [Storz, 1990a;b; STORZ, 1990c], an osmolyte-driven formation of reactive oxygen species may explain the observed induction of *ahpCF* and *dps*. To test a possible formation of reactive oxygen species by carbohydrate-mediated osmotic stress, the uncharged and non-fluorescent reactive oxygen species indicator dihydrorhodamin 123 was used. Dihydrorhodamin 123 passively diffuses across membranes where it is oxidised by intracellular $H_2O_2$ to cationic rhodamine 123, which exhibits fluorescence. Accordingly, the observed fluorescence of viable cells is proportional to the cytoplasmic $H_2O_2$ concentration [HENDERSON, 1993].

Dihydrorhodamine 123-labeled *E. coli* were grown in M9 medium only or stimulated with $H_2O_2$, sucrose, glucose, lactose or casamino acids (Figure 19). In the presence of $H_2O_2$ (600 µM), an approximately 2-fold higher fluorescence signal was detected compared to incubation of cells without stimuli (negative control). Exposure of cells to sucrose (400 mM) resulted in a 2-fold lower fluorescence signal than observed for $H_2O_2$ and was similar to the values observed for the negative control. In contrast, fluorescence signals obtained after stimulation of cells with glucose, lactose (50 mM and 400 mM each) or casamino acids (2%) did not differ from those observed for $H_2O_2$ (600 µM). Therefore, intracellular $H_2O_2$ formation is enhanced by carbohydrate or casamino acid catabolism but not by high medium osmolality alone. These results indicate that the OxyR-dependent induction of *ahpCF* and *dps* was directly mediated by osmotic stress.

# RESULTS

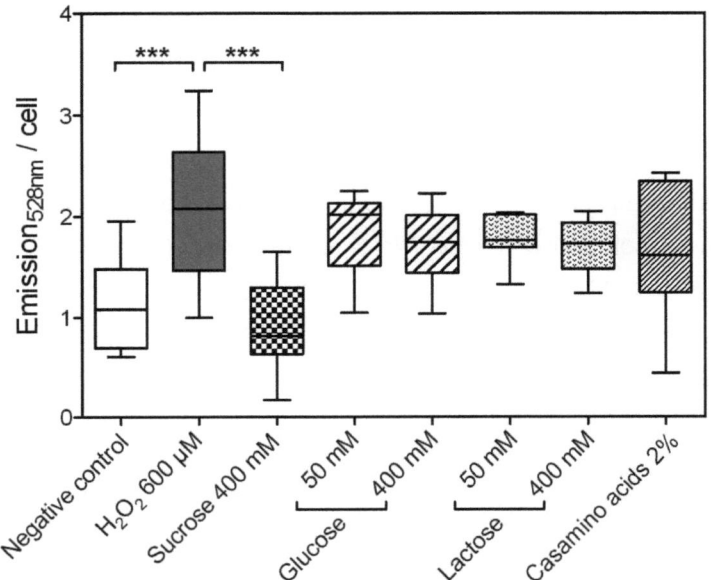

**Figure 19.** **No formation of intracellular H$_2$O$_2$ by osmotic stress.** Fluorescence of dihydrorhodamin 123-labelled *E. coli* after incubation without stimuli (negative control) or with H$_2$O$_2$, sucrose, glucose, lactose and casamino acids for 60 min at 37°C is shown. Data are expressed as medians and minima versus maxima (negative control, H$_2$O$_2$, n = 17; others, n = 7). Differences between treatment groups were calculated by the Kruskal-Wallis 1-Way AOV and the Dunn's multiple-comparison test. *, P < 0.05; **, P < 0.01; ***, P < 0.001.

## 4.3.1.4. Requirement of OxyR-regulated proteins at high osmolality

The observed upregulation of AhpF and Dps in the intestines of mice fed the lactose diet and the induction of the *ahpCF* and *dps* gene expression by osmolytes indicate a role of these OxyR-dependent proteins in the adaptation of *E. coli* to carbohydrate-induced osmotic stress. To confirm this assumption, growth of mutants lacking the *ahpCF* or *oxyR* genes was investigated under aerobic and anaerobic conditions (Table 3). On LB medium, all strains reached similar cell densities and had equal doubling times. To provide osmotically constant conditions, non-fermentable sucrose (400 mM and 700 mM) was added. The presence of 400 mM sucrose resulted in reduced cell densities of both mutants compared to the wild type (aerobic, by up to 40%; anaerobic, by up to 28%), but only *oxyR* deficient *E. coli* had longer doubling times (aerobic, 30%; anaerobic, 20%). For both mutants, a sucrose concentration of 700 mM affected the final cell densities more strongly: they decreased by up to 64% under aerobic and by up to 50% under anaerobic conditions in comparison to the wild type. As observed for the lower sucrose concentration (400 mM), only the doubling times of *E. coli* ΔoxyR increased in the presence of 700 mM sucrose by up to 35%, whereas there was no difference for *E. coli* lacking *ahpCF* (both compared to the wild type).

**Table 3.** Growth of wild type, *ahpCF* and *oxyR* deficient *E. coli* in the presence or absence of osmotic stress caused by non-fermentable sucrose [a].

| Medium [b] | | *E. coli* | *E. coli* Δ*ahpCF* | *E. coli* Δ*oxyR* |
|---|---|---|---|---|
| **AEROBIC GROWTH CONDITIONS** | | | | |
| LB | $OD_{600}$ after 24 h | 6.9 (5.5:8.0) | 6.3 (5.6:7.0) | 6.6 (6.3:6.7) |
| | Doubling time $t_d$ (min) | 33 (31:35) | 32 (27:36) | 39 (34:39) |
| LB, 400 mM sucrose | $OD_{600}$ after 24 h | 4.5 (3.8:4.5) | 2.6 (2.5:2.9) [c] | 2.7 (2.5:2.9) [c] |
| | Doubling time $t_d$ (min) | 36 (35:37) | 36 (34:40) | 47 (44:51) [c] |
| LB, 700 mM sucrose | $OD_{600}$ after 24 h | 2.8 (2.5:3.1) | 1.0 (1.0:1.2) [c] | 1.3 (1.3:1:3) [c] |
| | Doubling time $t_d$ (min) | 36 (33:37) | 40 (30:44) | 49 (44:56) [c] |
| **ANAEROBIC GROWTH CONDITIONS** | | | | |
| LB | $OD_{600}$ after 24 h | 0.85 (0.79:0.89) | 0.82 (0.76:0.9) | 0.78 (0.77:0.79) |
| | Doubling time $t_d$ (min) | 41 (35:45) | 41 (39:43) | 37 (36:42) |
| LB, 400 mM sucrose | $OD_{600}$ after 24 h | 0.43 (0.39:0.52) | 0.31 (0.30:0.33) [c] | 0.33 (0.30:0.36) [c] |
| | Doubling time $t_d$ (min) | 50 (39:54) | 54 (48:63) | 60 (59:68) [c] |
| LB, 700 mM sucrose | $OD_{600}$ after 24 h | 0.22 (0.19:0.27) | 0.11 (0.11:0.11) [c] | 0.11 (0.10:0.12) [c] |
| | Doubling time $t_d$ (min) | 66 (49:84) | 85 (61:98) | 89 (85:93) [c] |

[a] Data are expressed as medians and minima versus maxima, n = 4.
[b] Cultures were incubated on LB medium with or without sucrose at 180 rpm and 37°C.
[c] Differences between wild type and mutant *E. coli* were calculated by the U-test, P < 0.05.

# RESULTS

To ensure that the observed growth defects were indeed caused by the deletion of the *ahpCF* and *oxyR* genes, plasmids including the corresponding genes and the physiologically relevant promoters were constructed. Mutants containing these plasmids showed similar cell densities and doubling times as the wild type with the empty plasmid when grown on LB medium containing 400 mM or 700 mM sucrose (Table 4). The results demonstrate that *ahpCF* and *oxyR* improve the growth of *E. coli* under osmotic stress conditions.

Table 4. Growth of *ahpCF* and *oxyR* deficient *E. coli* containing complementing plasmids with the corresponding genes and promoters in comparison to the wild type with the empty vector under aerobic conditions [a].

| Medium [b] |  | *E. coli* pSU19 | *E. coli* Δ*ahpCF* pSU19-*ahpCF* | *E. coli* Δ*oxyR* pSU19-*oxyR* |
|---|---|---|---|---|
| LB | $OD_{600}$ after 24 h | 6.5 (6.1:6.8) | 6.5 (6.1:6.8) | 4.8 (4.6:5.0) [c] |
|  | Doubling time $t_d$ (min) | 37 (31:35) | 37 (31:35) | 35 (33:37) |
| LB, 400 mM sucrose | $OD_{600}$ after 24 h | 4.9 (4.1:5.1) | 5.3 (4.9:5.9) | 4.3 (3.8:4.7) |
|  | Doubling time $t_d$ (min) | 51 (49:54) | 57 (50:61) | 45 (42:49) |
| LB, 700 mM sucrose | $OD_{600}$ after 24 h | 4.2 (4.1:4.5) | 4.0 (3.8:4.1) | 4.1 (4.0:4.2) |
|  | Doubling time $t_d$ (min) | 97 (82:107) | 86 (77:100) | 91 (86:92) |

[a] Data are expressed as medians and minima versus maxima, n = 4.
[b] Cultures were incubated on LB medium with or without sucrose at 180 rpm and 37°C.
[c] Differences between wild type and mutant *E. coli* were calculated by the U-test, $P < 0.05$.

## 4.3.2. Analysis of proteins KduI and KduD

### 4.3.2.1. Induction of *kduI* and *kduD* gene expression by hexuronates

The proteomic data revealed the induction of KduD on the lactose diet and the repression of this protein and of KduI on the casein diet (see section 4.2.4.). Since the roles of these proteins in the *E. coli* metabolism are still obscure, the *kduD* gene expression in response to the diets and the intestinal substrates (glucose, fructose, lactose, galactose, casamino acids) was analysed using luciferase reporter gene assays (Table 5). To test the influence of host-endogenous substrates, small intestinal mucosal tissue was analysed. Galacturonate and glucuronate were included because they have been shown to induce the *kduI* and *kduD* gene expression in the plant pathogen *Erwinia chrysanthemi* [CONDEMINE, 1991].

Under aerobic conditions, no induction of the *kduD* promoter by any of the diets, the mucosal tissues, fructose, lactose or casaminoacids was observed. Only galactose, galacturonate and glucuronate increased the *kduD* promoter activity 3 to 4-fold. Under anaerobic conditions, only galacturonate and

**Table 5.** Induction of the *kduD* and *kduI* promoters in *E. coli* by the various mouse diets, small intestinal mucosal tissues, carbohydrates, casamino acids or hexuronates under aerobic and anaerobic growth conditions.

| Substrate [a] | Fold-change [b] | | | |
|---|---|---|---|---|
| | *kduD* | | *kduI* | |
| | aerobic | anaerobic | aerobic | anaerobic |
| Starch diet (1%) | 0.5 (0.5:0.6) | 3.4 (2.9:4.1) | n. d. | n. d. |
| Lactose diet (1%) | 1.0 (0.8:1.0) | 7.7 (5.8:8.9) | n. d. | n. d. |
| Casein diet (1%) | 0.2 (0.2:0.3) | 2.6 (2.2:4.0) | n. d. | n. d. |
| Mucosa of starch diet fed mice (1%) | 2.3 (1.9:3.4) | 0.1 (0.1:0.1) | n. d. | n. d. |
| Mucosa of lactose diet fed mice (1%) | 1.4 (1.3:1.5) | 0.1 (0.1:0.1) | n. d. | n. d. |
| Mucosa of casein diet fed mice (1%) | 1.7 (0.6:1.8) | 0.1 (0.1:0.1) | n. d. | n. d. |
| Fructose (50 mM) | 2.0 (1.5:2.8)) | 6.1 (4.1:25) | n. d. | n. d. |
| Lactose (25 mM) | 1.8 (1.2:2.2) | 3.2 (1.0:7.8) | n. d. | n. d. |
| Galactose (50 mM) | 4.0 (1.7: 4.2) [c] | n. d. | n. d. | n. d. |
| Casamino acids (50 mM) | 1.5 (1.2:2.8) | 9.8 (5.0:15) | n. d. | n. d. |
| Glucuronate (50 mM) | 3.2 (2.2:4.9) [c] | 20 (17:28) [d] | 7.0 (6.0:10) [d] | 54 (41:57) [e] |
| Galacturonate (50 mM) | 3.3 (2.3:3.5) [c] | 9.4 (6.9:13) [c] | 11 (5.0:14) [e] | 19 (16:21) [c] |

[a] Cultures were grown on M9 minimal medium containing pulverised mouse diets, scratched small intestinal mucosa or dietary components under aerobic or anaerobic conditions and shaking at 220 and 120 rpm, respectively, at 37°C for 16 h.
[b] Relative luminescence data for *E. coli* carrying pkduDp::luxAB or pkduIp::luxAB are shown. For both aerobic and anaerobic growth conditions, luciferase activity was normalised to values determined for cells grown on glucose (50 mM). Data are expressed as medians and interquartile ranges (aerobic, n = 5-7; anaerobic, n = 4-7).
[c,d,e] Kruskal-Wallis 1-Way AOV and Dunn's multiple-comparison test were used for calculations. [c], $P < 0.05$; [d], $P < 0.01$; [e], $P < 0.001$.
n. d. = not determined

glucuronate increased *kduD* expression 9-fold and 20-fold, respectively (Table 5). Since *E. coli* K-12 is unable to grow on galactose-containing minimal medium in an anoxic environment [MUIR, 1985], it was not possible to analyse the effect of galactose on the *kduD* expression under these conditions.

Since KduI was repressed in intestinal *E. coli* of casein-fed mice, only substances that induced the *kdud* gene expression were tested for their effect on the *kduI* promoter activity. In the presence of galacturonate and glucuronate the *kduI* promoter activity increased under both aerobic and anaerobic conditions (Table 5). As observed for *kduD*, anaerobic cultivation resulted in higher *kduI* expression levels compared to aerobic cultivation (galacturonate, 19-fold versus 11-fold; glucuronate, 54-fold versus 7-fold).

Under all experimental conditions tested, expression of *kduI* was about 2 to 3-fold higher than that of *kduD*. Furthermore, both promoters were induced more strongly under anaerobic than under aerobic conditions. In an anoxic environment, glucuronate led to higher promoter activities than galacturonate, whereas equal values were detected for aerobically grown *E. coli*. In conclusion, these experiments demonstrate the influence of galacturonate and glucuronate on the *kduD* and *kduI* promoter activity.

### 4.3.2.2. Repression of *uxaCA*, *uxaB* and *uxuAB* expression by osmotic stress

The induction of the *kduI* and *kduD* gene expression by galacturonate and glucuronate suggests a contribution of the corresponding proteins in the conversion of these hexuronates. However, until now, there is no evidence for a role of KduI and KduD in hexuronate breakdown. In *E. coli*, galacturonate and glucuronate conversion is normally catalysed by uronate isomerase (UxaC), altronate oxidoreductase (UxaB) or mannonate oxidoreductase (UxuB), and altronate dehydratase (UxaA) or mannonate dehydratase (UxuA)

[RODIONOV, 2000]. Interestingly, the promoter of the *uxuAB* operon contains an OxyR-binding site and *uxuA* expression is higher in the absence of *oxyR* [ZHENG, 2001b]. Based on this observation and the sensitivity of OxyR-dependent proteins against osmotic stress (described in section 4.3.1.), the osmotic effect of lactose was hypothesised to repress the expression of the regular hexuronate degrading enzymes UxaABC and UxuAB in an OxyR-dependent manner. Under such conditions, KduI and KduD might compensate for the metabolic function of these enzymes.

To test this hypothesis, the gene expression of *uxaCA*, *uxaB* and *uxuAB* in response to galacturonate and glucuronate was analysed using luciferase reporter gene assays (Figure 20). These analyses revealed that incubation of *E. coli* with galacturonate and glucuronate increased the promoter activity of *uxaCA*, *uxaB* and *uxuAB* compared to incubation without hexuronates (negative control) under both aerobic and anaerobic growth conditions. To investigate a range of osmolalities, 25 to 400 mM non-fermentable sucrose was added to galacturonate- or glucuronate-containing medium. With increasing sucrose concentrations, the *uxaCA*, *uxaB* and *uxuAB* promoter activity decreased by up to approximalety 60% under both aerobic and anaerobic conditions. The same effect was observed in the presence of the osmolyte NaCl (400 mM), which supports the notion that osmotic stress is responsible for the observed *uxaCA*, *uxaB* and *uxuAB* repression (data not shown). Addition of $H_2O_2$ (300 µM), a known activator of OxyR, resulted in a decreased activity of all promoters in an aerobic (Figure 20) but not in an anaerobic environment, whose reducing conditions destabilised the added $H_2O_2$ (data not shown). These results demonstrate that not only $H_2O_2$ but also osmolytes such as carbohydrates and NaCl repressed the *uxaCA*, *uxaB* and *uxaAB* promoters.

To clarify the role of OxyR in *uxaCA*, *uxaB* and *uxuAB* expression, luciferase reporter gene assays were performed in *oxyR* deficient *E. coli* (Figure 21). In

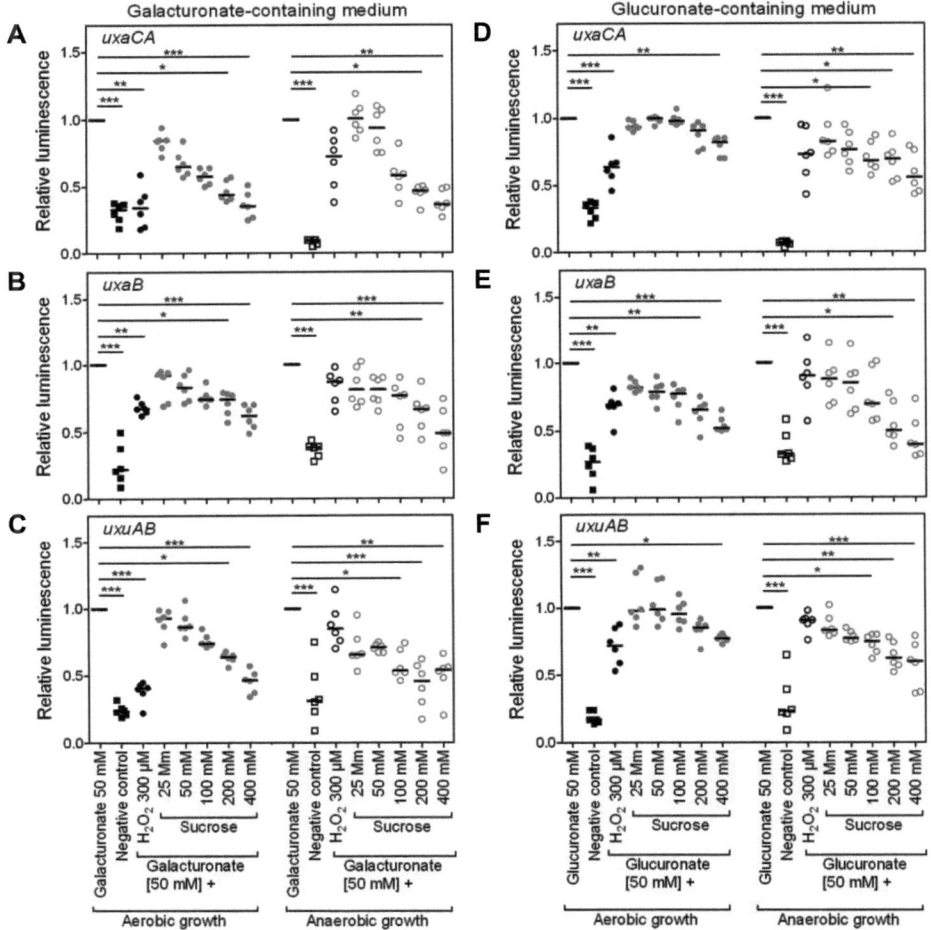

**Figure 20. Repression of the *uxaCA*, *uxaB* and *uxuAB* promoters by osmolytes.** *E. coli* carrying p*uxaCAp::luxAB* (**A, D**), p*uxaBp::luxAB* (**B, E**) or p*uxuABp::luxAB* (**C, F**) were incubated in M9 minimal medium without substrate (negative control), with galacturonate or glucuronate (50 mM each) and with $H_2O_2$ (300 µM) or various concentrations of sucrose for 90 min under aerobic (filled symbols) or anaerobic (open symbols) conditions at 37°C. Relative luminescence data are shown. Luciferase activity was normalised to values determined for cells grown on 50 mM galacturonate and glucuronate, respectively. Data are expressed as medians (n = 6). Differences between treatment groups were calculated by the Kruskal-Wallis 1-Way AOV and the Dunn's multiple-comparison test. *, $P < 0.05$; **, $P < 0.01$; ***, $P < 0.001$.

# RESULTS

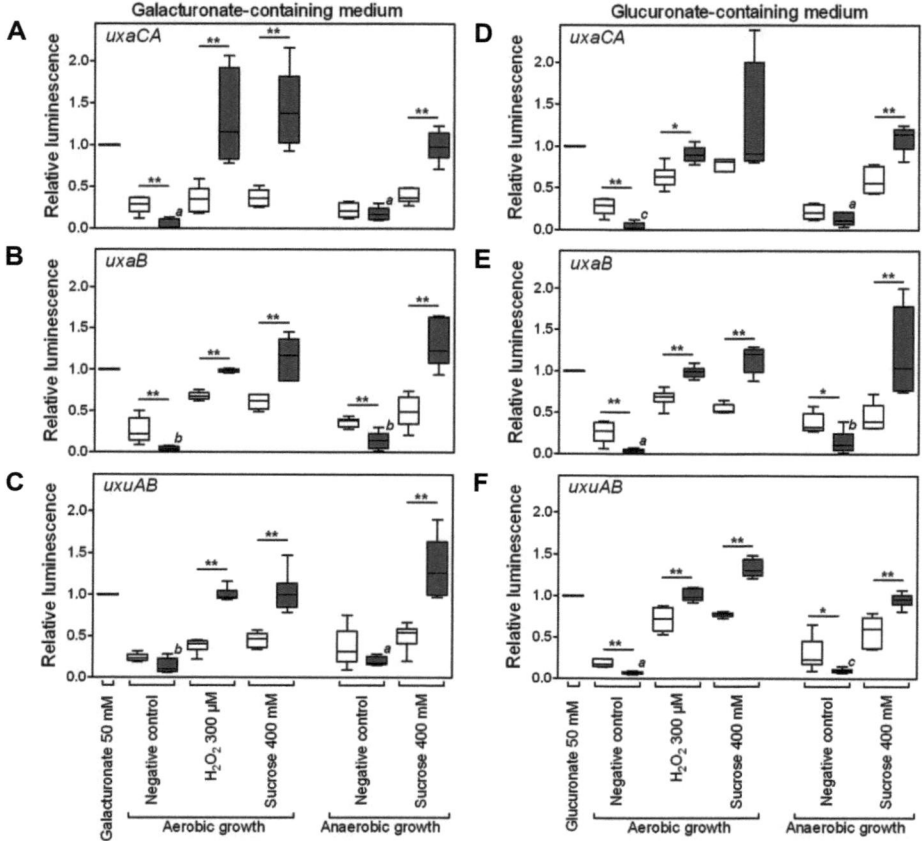

**Figure 21. OxyR-dependency of the *uxaCA*, *uxaB* and *uxuAB* gene expression.** *E. coli* (white bars) and *E. coli* ΔoxyR (grey bars) carrying p*uxaCAp::luxAB* (**A, D**), p*uxaBp::luxAB* (**B, E**) or p*uxuABp::luxAB* (**C, F**) were incubated in M9 minimal medium without substrate (negative control), with galacturonate or glucuronate (50 mM each) and $H_2O_2$ (300 µM) or sucrose (400 mM) for 90 min under aerobic and anaerobic conditions at 37°C. Relative luminescence data are shown. Luciferase activity was normalised to values determined for cells grown on 50 mM galacturonate and glucuronate, respectively. Data are expressed as medians and minima versus maxima (n = 6). The Kruskal-Wallis 1-Way AOV and Dunn's multiple-comparison test were used for calculations of values obtained from wild type *E. coli*. The U-test was applied to compare wild type and *E. coli* ΔoxyR. *[a], $P < 0.05$; **[b], $P < 0.01$; ***[c], $P < 0.001$.

contrast to the observations in the wild type with a functional *oxyR*, aerobic incubation on galacturonate- or glucuronate-containing medium with 300 µM $H_2O_2$ did not diminish the *uxaCA*, *uxaB* and *uxuAB* promoter activity. This suggests that OxyR acts as a repressor on the expression of these genes. The same effect was observed in the presence of 400 mM sucrose: in the absence of *oxyR*, higher activities of all promoters were detected compared to the wild type under both aerobic and anaerobic growth conditions. Therefore, carbohydrate-induced osmotic stress led to the OxyR-mediated repression of gene expression of *E. coli*'s regular hexuronate degrading enzymes UxaABC and UxuAB.

### 4.3.2.3. No repression of *kduI* and *kduD* gene expression by osmotic stress

Since KduI and KduD are suggested to be involved in hexuronate conversion under conditions, in which *uxaCA*, *uxaB* and *uxuAB* are repressed, the dependence of the *kduI* and *kduD* gene expression on OxyR and their expression in response to osmotic stress was investigated. For this purpose, luciferase reporter gene assays using the *kduI* and *kduD* promoters were performed in *oxyR*-deficient *E. coli* and compared with results obtained from the wild type with a functional *oxyR* (Figure 22). Under aerobic and anaerobic conditions, incubation with galacturonate or glucuronate increased the *kduI* and *kduD* promoter activity to a similar extent in wild type and Δ*oxyR* cells. In the presence of glucuronate, addition of 400 mM sucrose did not change the *kduI* and *kduD* promoter activity compared to incubation without osmotic stress in both wild type and *E. coli* lacking the *oxyR* gene. However, in galacturonate-containing medium, the presence of sucrose promoted the *kduI* and *kduD* gene expression in *E. coli* Δ*oxyR*. These results suggest that the hexuronate-driven induction of *kduI* and *kduD* is not repressed by OxyR. That *kduI* and *kduD* gene expression is also induced in the presence of osmotic stress supports the hypothesis that KduI and KduD compensates for the function of UxaABC and UxuAB, whose expression is repressed under such conditions.

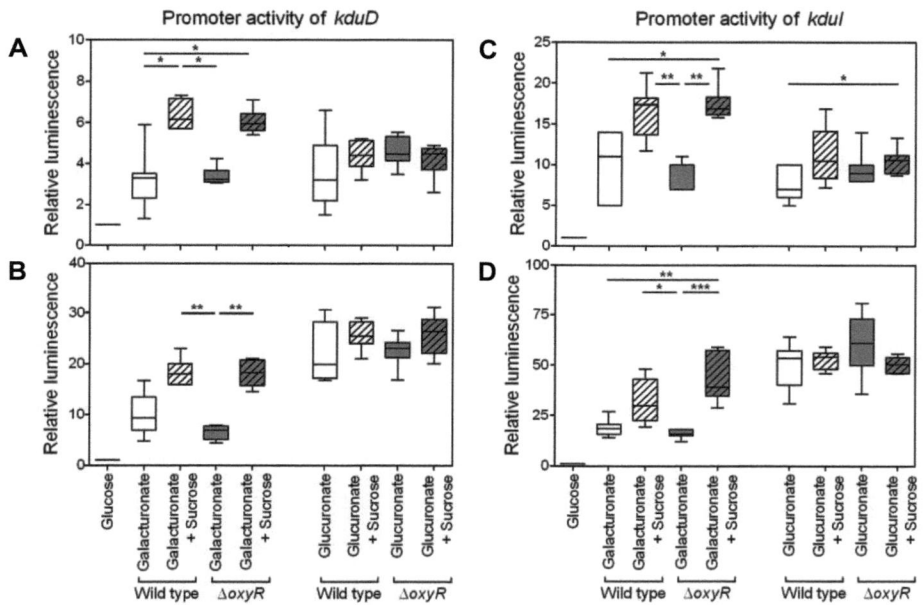

**Figure 22.** Effect of osmotic stress on hexuronate-induced expression of *kduI* and *kduD*. Comparison of relative luminescence of *E. coli* (white bars) and *E. coli* Δ*oxyR* (grey bars) carrying p*kduDp::luxAB* (A, B) and p*kduIp::luxAB* (C, D) in response to incubation in M9 minimal medium containing 50 mM glucuronate and galacturonate, respectively, with or without 400 mM sucrose for 16 h under aerobic (**A, C**) or anaerobic (**B, D**) conditions at 37°C. Luciferase activity was normalised to values determined for cells grown on 50 mM glucose. Data are expressed as medians and minima versus maxima (n = 4-7). Differences between treatment groups were calculated by the Kruskal-Wallis 1-Way AOV and the Dunn's multiple-comparison test. *, $P < 0.05$; **, $P < 0.01$; ***, $P < 0.001$.

### 4.3.2.4. Promotion of hexuronate conversion by KduI and KduD

To investigate the role of KduI and KduD in galacturonate and glucuronate breakdown, the conversion of these hexuronates by cell-free extracts of

E. coli overexpressing KduI, KduD or both was examined (Figure 23). For this purpose, E. coli cell-free extracts were incubated with galacturonate or glucuronate (10 mM each). Hexuronate concentrations were determined enzymatically using uronate dehydrogenase, which converts D-galacturonate and D-glucuronate to D-galactarate and D-glucarate, respectively [MOON, 2009]. The conversion of galacturonate and glucuronate was higher in cell-free extracts of E. coli overexpressing both, KduI and KduD or only KduI compared to extracts of cells containing the empty vector or overexpressing KduD alone. Accordingly, the specific activities of cell-free extracts containing

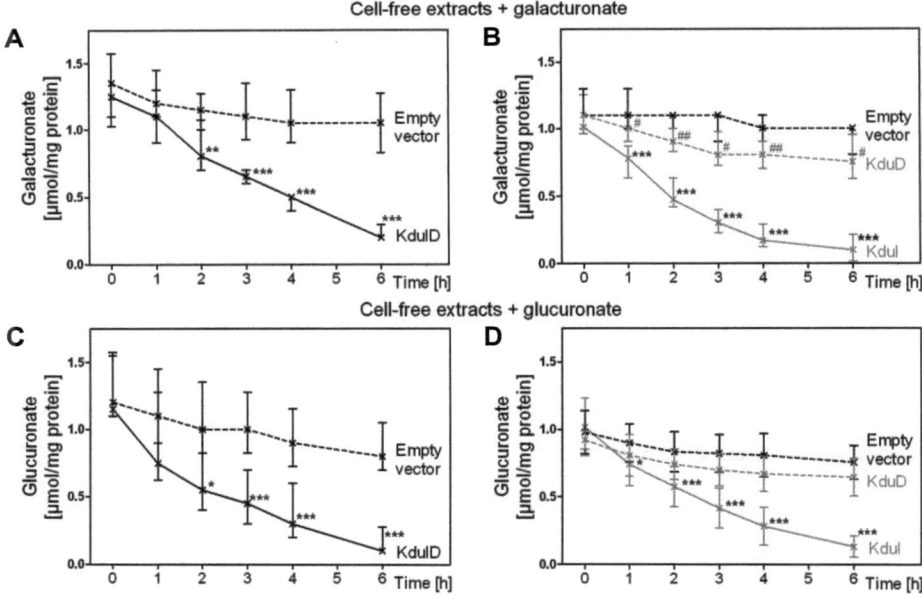

Figure 23. Breakdown of galacturonate and glucuronate by KduI and KduD. Cell-free extracts of E. coli overexpressing KduI, KduD or both were incubated with 10 mM galacturonate or 10 mM glucuronate at 37°C. Hexuronate concentration at specific time points was determined enzymatically. E. coli strains JM109 (A, C) or KRX (B, D) were selected according to the orientation of the cloned genes. Data are expressed as medians and interquartile ranges (n = 11-12). Differences between extracts of negative controls and overexpressed proteins (*, KduI; #, KduD) were calculated by the U-test. *#, $P < 0.05$; **, $P < 0.01$; ***, $P < 0.001$.

KduI and KduD or KduI alone were 2 to 5-fold higher than those of cell-free extracts containing the empty vector or overexpressing only KduD (Table III, APPENDIX II). These results indicate the ability of KduI to facilitate the conversion of galacturonate and glucuronate.

### 4.3.2.5. Requirement of KduID for growth on hexuronates at high osmolality

To elucidate whether the observed conversion of hexuronates in the presence of KduI is of biological relevance under osmotic stress conditions, growth of wild type and *kduID* deficient *E. coli* was monitored in an aerobic and anaerobic environment (Table 6, Figure 24). When cultivated in the presence of galacturonate or glucuronate as sole energy source, both strains grew to similar maximal cell densities and reached equal doubling times. In the presence of osmotic stress caused by non-fermentable sucrose (200 mM on glucuronate, 400 mM on galacturonate medium), which repressed the expression of *uxaCA*, *uxaB* and *uxuAB*, *kduID* deficient *E. coli* had 1.5 to 2-fold longer doubling times than the wild type, but the maximal cell densities did not differ. Sucrose concentrations of 400 mM on glucuronate- and 700 mM on galacturonate-containing medium resulted in 1.5 to 2-fold longer doubling times as well as diminished maximal cell densities of *E. coli* Δ*kduID* compared to the wild type. However, this effect was much more pronounced in an aerobic than in an anaerobic environment: maximal cell densities were reduced by up to 90% under aerobic and by up to 30% under anaerobic growth conditions.

# RESULTS

**Table 6.** Growth behaviour of wild type and *kdulD* deficient *E. coli* on galacturonate- or glucuronate-containing medium with or without osmotic stress caused by non-fermentable sucrose [a].

| Medium [b] | $OD_{600}$ max. | | Doubling time $t_d$ (min) | |
|---|---|---|---|---|
| | *E. coli* | *E. coli* Δ*kdulD* | *E. coli* | *E. coli* Δ*kdulD* |
| **AEROBIC GROWTH CONDITIONS** | | | | |
| Glucuronate | 4.8 (4.7:4.9) | 4.8 (4.6:5.0) | 75 (74:80) | 80 (78:83) [c] |
| Glucuronate, 200 mM sucrose | 4.2 (4.0:4.5) | 3.8 (3.5:4.0) [d] | 86 (81:98) | 182 (179:189) [d] |
| Glucuronate, 400 mM sucrose | 1.4 (1.2:2.0) | 0.3 (0.3:0.7) [d] | 141 (128:156) | 270 (221:326) [d] |
| Galacturonate | 5.9 (5.8:6.1) | 5.9 (5.6:6.1) | 75 (71:76) | 73 (72:103) |
| Galacturonate, 400 mM sucrose | 4.0 (3.8:4.3) | 4.0 (3.8:4.3) | 114 (109:120) | 252 (200:278) [d] |
| Galacturonate, 700 mM sucrose | 1.9 (1.0:2.7) | 0.22 (0.14:0.56) [d] | 159 (133:186) | 382 (284:522) [d] |
| **ANAEROBIC GROWTH CONDITIONS** | | | | |
| Glucuronate | 0.71 (0.62:0.76) | 0.7 (0.61:0.89) | 114 (99:145) | 118 (101:165) |
| Glucuronate, 200 mM sucrose | 0.43 (0.41:0.5) | 0.35 (0.3:0.36) [d] | 202 (171:225) | 281 (241:453) [d] |
| Glucuronate, 400 mM sucrose | 0.35 (0.35:0.4) | 0.25 (0.21:0.29) [d] | 237 (223:377) | 382 (358:441) [c] |
| Galacturonate | 0.76 (0.58:0.85) | 0.69 (0.65:0.82) | 133 (129:153) | 125 (122:153) |
| Galacturonate, 400 mM sucrose | 0.36 (0.35:0.49) | 0.34 (0.34:0.36) | 183 (151:265) | 309 (270:380) [d] |
| Galacturonate, 700 mM sucrose | 0.35 (0.3:0.37) | 0.24 (0.22:0.28) [d] | 309 (271:340) | 424 (402:537) [d] |

[a] Data are expressed as medians and minima versus maxima (aerobic, n = 6; anaerobic, n = 5).
[b] Cultures were incubated on M9 minimal medium containing galacturonate or glucuronate (50 mM each) with or without sucrose at 37°C. Aerobic cultures were shaken at 218 rpm.
[c, d] Differences between wild type and mutant *E. coli* on the same medium were calculated by the U-test.
[c], $P < 0.05$. [d], $P < 0.01$.

# RESULTS

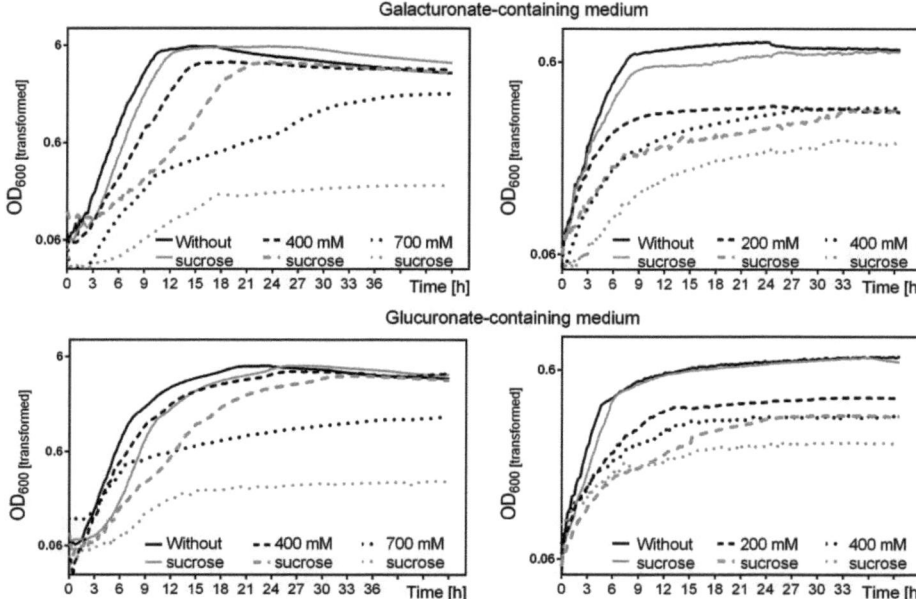

**Figure 24. Growth retardation of *E. coli* Δ*kduID* by osmotic stress.** *E. coli* and *E. coli* Δ*kduID* were incubated in M9 minimal medium containing 50 mM galacturonate (A, B) or 50 mM glucuronate (C, D) with or without sucrose at 37°C. Black, dark green and dark red lines: *E. coli*; grey, light green and light red lines: *E. coli* Δ*kduID*. **(A, C)** Aerobic conditions; **(B, D)** anaerobic conditions. Cell densities were determined at 600 nm in intervals of 15 min. Data represent medians of 6 replicates for aerobic and 5 replicates for anaerobic conditions.

To ensure that the observed growth defects were caused by the deletion of the *kduID* genes, *E. coli* Δ*kduID* were transformed with plasmids containing the corresponding genes and physiologically relevant promoters. Growth was monitored on galacturonate- and glucuronate-containing medium in the presence of 400 mM sucrose under aerobic conditions (Table 7). Mutants containing these plasmids showed similar maximal cell density and doubling times as the wild type with the empty vector. The results illustrate that KduI and KduD facilitate the growth of *E. coli* on galacturonate and glucuronate under osmotic stress conditions.

**Table 7.** Growth behaviour of *kduID* deficient *E. coli* containing complementing plasmids in comparison to the wild type with the empty vector under aerobic conditions [a].

| Medium [b] | | E. coli pSU19 | E. coli ΔkduID pSU19-kduID |
|---|---|---|---|
| Glucuronate | OD$_{600}$ after 24 h | 3.6 (3.4:3.8) | 4.2 (4.0:4.4) [c] |
| | Doubling time $t_d$ (min) | 89 (75:113) | 82 (80:93) |
| Glucuronate, 400 mM sucrose | OD$_{600}$ after 24 h | 3.4 (3.0:3.8) | 4.2 (3.8:4.4) [c] |
| | Doubling time $t_d$ (min) | 218 (183:226) | 237 (195:271) |
| Galacturonate | OD$_{600}$ after 24 h | 5.4 (5.3:5.4) | 5.3 (5.2:5.4) |
| | Doubling time $t_d$ (min) | 78 (74:82) | 82 (75:86) |
| Galacturonate, 400 mM sucrose | OD$_{600}$ after 24 h | 4.1 (4.0:4.3) | 5.2 (3.7:6.3) |
| | Doubling time $t_d$ (min) | 122 (115:125) | 134 (94:208) |

[a] Data are expressed as medians and minima versus maxima (n = 6).
[b] Cultures were incubated on M9 minimal medium containing galacturonate or glucuronate (50 mM each) with or without 400 mM sucrose at 218 rpm and 37°C.
[c] Differences between the wild type with the empty vector and *kduID* deficient *E. coli* with the complementing plasmids on the same medium were calculated by the U-test. P < 0.05.

### 4.3.2.6. Origin of hexuronates in the mouse intestine

The above described experiments demonstrate a role of KduI and KduD in galacturonate and glucuronate metabolism under conditions of osmotic stress, which repressed *E. coli*'s regular hexuronate degrading enzymes UxaABC and UxuAB. Since KduD was upregulated in intestinal *E. coli* of lactose-fed mice, the question arose whether hexuronates were available at higher levels in the gut of these mice and whether they possibly originate from dietary lactose. For this purpose, the hexuronate concentrations in the intestinal contents were determined enzymatically using uronate dehydrogenase. Since this enzyme cannot distinguish between galacturonate and glucuronate, the measured concentrations are given as hexuronates in the following (see section 4.3.2.4.) [MOON, 2009]. The hexuronate concentrations in small intestinal, caecal and colonic contents of mice fed the lactose diet were higher than in those of mice fed the starch or the casein diet (Table 8). The maximal concentration of 104 μg hexuronates per g [ww] corresponds to 0.48 mM galacturonate or 0.44 mM glucuronate.

**Table 8.** Concentrations of galacturonate and glucuronate in the intestinal contents of mice fed the starch, the lactose or the casein diet determined with an uronate dehydrogenase assay [a].

|  | Starch diet | Lactose diet | Casein diet |
|---|---|---|---|
| Small intestine | 40 (17:64) | 104 (61:130) *,# | 42 (0:91) |
| Caecum | 14 (11:19) | 28 (15:38) *** | 0 (0:0) |
| Colon | 5 (0:16) | 14 (11:20) *** | 0 (0:0) |

[a] Data are given in µg per g [ww] and expressed as medians and interquartile ranges, n = 10-11.
* Differences between feeding groups were calculated by the Kruskal-Wallis 1-Way AOV and the Dunn's multiple-comparison test.
#, lactose versus starch diet, $P < 0.05$. *, lactose versus casein diet. *, $P < 0.05$; ***, $P < 0.001$.

To investigate whether hexuronates possibly originate from dietary lactose, the intracellular hexuronate concentration of *E. coli* grown in the presence of lactose (25 mM) or glucose (50 mM) were measured at selected time points under aerobic conditions. It was not possible to perform this experiment under anaerobic conditions because of the insufficient cell density. Intracellular hexuronate was only measured during growth on lactose but not on glucose (Figure 25). The highest hexuronate concentration was determined during the exponential growth phase (138 ng hexuronates per mg protein). Based on the assumption that 1 mg protein corresponds to a cytoplasmic volume of 2 µl [STOCK, 1977; SMITH, 1999; KAWANO, 2000], 138 ng hexuronates per mg protein corresponds to a concentration of 0.32 mM galacturonate or 0.3 mM glucuronate in the bacterial cytoplasm. In the stationary phase of *E. coli* grown on lactose, the hexuronate concentrations decreased continuously (2.4 ng hexuronates per mg protein). This transient generation and disappearance of hexuronates during lactose metabolism revealed that these hexuronates are catabolised by *E. coli* and serve as energy source.

In conclusion, these experiments demonstrate that hexuronates are present to a larger extent in the small intestine of lactose-fed mice compared to starch- or casein-fed mice and that intracellular hexuronates are generated during growth of *E. coli* on lactose. Therefore, dietary lactose seems to affect hexuronate availability for intestinal *E. coli*. Nevertheless, it was not possible to clearly identify potential precursors of the detected intracellular hexuronates.

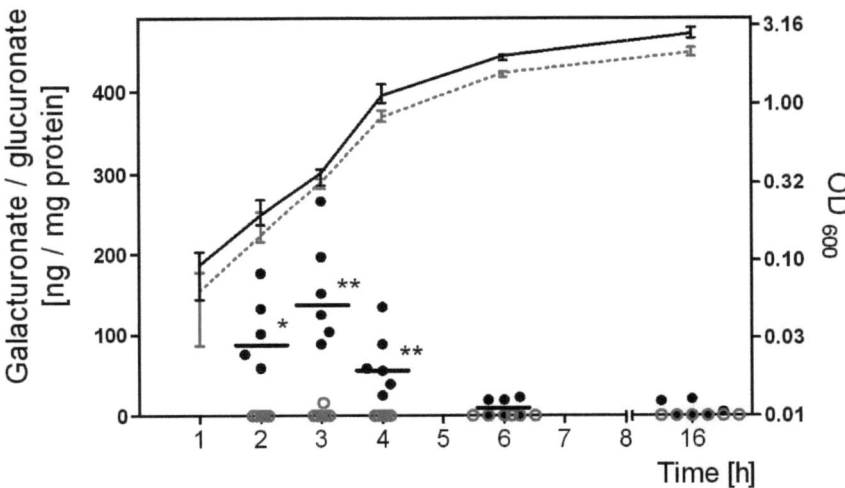

**Figure 25.** **Generation of intracellular hexuronate in exponential phase of *E. coli* during growth on lactose.** *E. coli* was cultivated in M9 minimal medium containing 50 mM glucose (grey) or 25 mM lactose (black) under aerobic conditions. Cell densities (lines) were determined at 600 nm. Intracellular galacturonate and glucuronate concentrations (dots) were determined using an uronate dehydrogenase assay. Data are expressed as medians and minima to maxima (n = 6). Differences between treatment groups were calculated by the U-test. *, $P < 0.05$; **, $P < 0.01$.

## 5. Discussion

The response of intestinal bacteria to dietary alterations has been studied extensively at the population level, but owing to the complexity of the commensal microbiota and the high inter-individual variability, little knowledge exits about the events occurring at the cellular level. Therefore, the objective of this work was to identify mechanisms that enable intestinal bacteria to adapt to nutritional factors using a simplified animal model. For this purpose, gnotobiotic mice monoassociated with *E. coli* MG1655 were fed either one of three diets: a diet rich in starch, a diet rich in non-digestible lactose or a diet rich in casein. The proteomes of *E. coli* recovered from small intestine and caecum of mice fed the lactose or the casein diet were compared with those of *E. coli* obtained from mice fed the starch diet and differentially expressed bacterial proteins were identified. Selected proteins were further characterised *in vitro* for their possible roles in the bacterial adaptation to the various diets.

### 5.1. Gnotobiotic mice – a useful model to analyse the effect of dietary factors

Feeding mice different diets was expected to lead to changes in the intestinal substrate availability. In the small intestine, dietary starch is almost completely digested [WEURDING, 2001; WILFART, 2007] and the resulting malto-oligosaccharides and glucose are rapidly absorbed. Therefore, and because of the absence of starch-degrading enzymes in *E. coli* [FREUNDLIEB, 1986], feeding mice the diet rich in starch (43% starch [wt/wt]) was supposed to minimise the availability of dietary substrates for intestinal *E. coli* and to promote the utilisation of host-endogenous substrates, such as mucosal glycoproteins and glycolipids [PEEKHAUS, 1998]. Hence, this diet served as a control. To analyse carbohydrate-induced changes in bacterial metabolism, mice were fed the lactose diet (10% lactose [wt/wt]). As lactose is not digestible by adult rodents [TROELSEN, 1992; KAUR, 2006], this carbohydrate was assumed to be completely available for intestinal *E. coli*. The high protein concentration present in the casein diet (60% casein [wt/wt]) was supposed to be digested and absorbed by the host only incompletely. Therefore, dietary peptides and amino acids would be partially available to intestinal bacteria, thereby enabling the analysis of the influence of dietary proteins on the metabolism of *E. coli*.

To investigate whether the various diets did indeed affect substrate availability in the mouse gut, the intestinal concentrations of sucrose, fructose, glucose, lactose, proteins and amino acids were determined. Sucrose and fructose were present at concentrations of less than 1 mM on any of the diets and are therefore not expected to significantly affect the growth of intestinal *E. coli*. Intestinal glucose originates from the cleavage of dietary sucrose and starch by host digestive enzymes. Both, the starch and the lactose diet, contained high proportions of starch (33 to 43% [wt/wt]), which was the main carbohydrate source on these diets. Accordingly, the highest glucose concentrations were detected in the small intestine of mice fed the starch and the lactose diet, whereas the glucose concentrations on the casein diet (3% starch [wt/wt]) were below 0.5 mM. The low concentrations of sucrose and fructose on any of the diets and the disappearance of glucose in the caecum and colon of starch- and lactose-fed mice correspond with the fact that nutrient-digestion and absorption happen preferentially in the ileum [FERRARIS, 2001]. Because of the indigestibility of dietary lactose by adult rodents [TROELSEN, 1992; KAUR, 2006], lactose was available for intestinal *E. coli* of mice fed the lactose diet by up to 10 mM, which was sufficient for the bacterial metabolism.

In contrast to the clear distribution of intestinal carbohydrates, proteins and amino acids were detected at a low level in the intestines of mice fed any of the diets. In the mouse intestine, not only dietary proteins but also host-endogenous proteins, such as digestive enzymes or proteins from desquamated epithelial cells, are present. Accordingly, disappearance of intestinal proteins and amino acids, as observed for digestible carbohydrates, cannot be expected. However, mice fed the casein diet displayed two-fold higher amino acid concentrations than mice fed the starch diet or the lactose diet. These conditions were considered to sufficiently mimic growth of *E. coli* on casein-derived substrates and support the validity of this model.

To ensure that the substrate availability did not affect intestinal colonisation, *E. coli* cell numbers were determined. The cell density in the small intestine ($\log_{10}$ CFU: 9 to 10) was lower than those determined for caecum ($\log_{10}$ CFU: 10 to 12) and colon ($\log_{10}$ CFU: 10 to 11). These numbers are comparable with results of other studies in gnotobiotic mice fed a standard chow and monoassociated with *E. coli* ($\log_{10}$ CFU: small intestine, 8; caecum, 11 to 12; colon, 10.5 to 11) [VOGEL-SCHEEL, 2010; SCHUMANN, 2012]. Interestingly, the cell numbers for the small intestine determined in the actual study are 1 to 2 log units higher than those of mice fed the standard chow. This may reflect the high availability of simple carbohydrates, peptides and amino acids in the intestines of mice fed the semisynthetic diets used in the actual study in contrast to the plant polysaccharide-rich standard chow. Moreover, feeding mice the lactose diet promoted the growth of intestinal *E. coli* as obvious from higher small intestinal and caecal *E. coli* counts compared to those of mice fed the starch or the casein diet. However, due to the monoassociated status of the animals, it remains unclear if this higher cell density is of biological relevance. These results indicate that gnotobiotic mice fed the three diets were a useful model for investigating the influence of dietary factors on the proteome of intestinal bacteria.

## 5.2. Adaptation of intestinal *E. coli* to dietary factors

The proteomes of intestinal *E. coli* obtained from mice fed the lactose and the casein diet were compared with those of *E. coli* recovered from mice fed the starch diet. This comparison revealed more than one hundred differentially expressed proteins, indicating the high capacity of *E. coli* to adapt to the ingested host diets. The adaptation of this bacterium to the metabolism of lactose and amino acids could be verified: *E. coli* obtained from mice fed the lactose diet induced Leloir pathway enzymes required for the utilisation of lactose [HOLDEN, 2003] and enzymes involved in amino acid biosynthetic processes, such as glutamate dehydrogenase (GdhA) and carbamoyl-phosphate synthase (CarA). The latter enzymes catalyse the biosynthesis of glutamate [SAKAMOTO, 1975] and arginine and pyrimidine nucleotides [THODEN, 1999], respectively. Their induction indicates a deficiency of amino acids and pyrimidine in the gut of lactose-fed mice. A shortage of intestinal nucleosides in mice monoassociated with *E. coli* has previously been demonstrated by VOGEL-SCHEEL *et al.*, who observed that key enzymes of *E. coli*'s purine and pyrimidine biosynthesis (*purC*, *pyrBI*) are essential for the colonisation of the mouse gut [VOGEL-SCHEEL, 2010]. In contrast, enzymes involved in peptide and amino acid catabolic processes, such as the glutamine degrading glutaminase 1 (GlsA1) [BROWN, 2008], were downregulated in *E. coli* obtained from mice fed the lactose diet.

Although the proteomic data clearly demonstrated the adaptation of intestinal *E. coli* to the lactose-rich diet, an apparent induction of enzymes required for peptide and amino acid breakdown was hardly detectable on the casein diet. The number and the fold-change of differentially expressed proteins of *E. coli* recovered from mice fed the casein diet was lower compared to those obtained from *E. coli* of mice fed the lactose diet (both in comparison to the proteome of mice fed the starch diet). In contrast to the catabolism of simple carbohydrates, such as glucose and lactose, the degradation of the large variety of different amino acids originating from casein digestion involves numerous specific enzymes. These enzymes undergo many subtle and less obvious changes, which are probably not detectable with the applied proteomic approach.

## 5.3. OxyR-dependent stress-response proteins are crucial under osmotic stress

Interestingly, the majority of proteins involved in *E. coli*'s cell redox homeostasis and stress response were upregulated on the lactose diet and downregulated on the casein diet. The genes encoding some of these upregulated proteins, such as *ahpCF*, *dps* and *fur*, belong to the *oxyR* regulon. The corresponding protein encoded by *oxyR* belongs to the large family of bacterial DNA-binding transcriptional regulators [STORZ, 1990b]. OxyR negatively autoregulates its own expression [CHRISTMAN, 1989; TAO, 1991] and acts as transcriptional inhibitor or enhancer of a large variety of genes, whose corresponding proteins are involved in peroxide metabolism, protection against peroxides and redox balance. In addition, OxyR induces the expression of important transcriptional regulators, such as *oxyS* [STORZ, 1990a; ZHENG, 2001a; ZHENG, 2001b]. It has been shown that oxidative stress exerted by $H_2O_2$ leads to the activation of OxyR via formation of reversible disulfide bonds between Cys199 and Cys208. This in turn provokes a conformational change and induces the transcription of the above mentioned antioxidant defence genes [STORZ, 1990b; TOLEDANO, 1994; LEE, 2004]. Therefore, OxyR acts as a sensor of oxidative stress and is essential for *E. coli*'s defence against it [STORZ, 1990a; STORZ, 1999]. The proteomic analysis did not reveal upregulation of other typical members of this regulon, such as superoxide dismutase (SodB), catalase peroxidase (KatG) and thioredoxin reductase (TrxB), possibly because of the inability to detect them. In total, approximately 60% of the detected differentially expressed protein spots could only be clearly identified by mass spectrometry. Therefore, it cannot be excluded that these proteins are among the unidentified protein spots.

Since the oxygen pressure in the intestine is low, it is doubtful that the observed upregulation of the OxyR-dependent proteins AhpF, Dps and Fur was a result of oxidative stress. It is also difficult to imagine that the presence of lactose promoted oxidative stress in the mouse gut while the casein diet appeared to prevent such stress. Since the *oxyR* regulon not only controls genes facilitating the adaptation to oxidative stress, but also the adaptation to heat shock, pH and salt stress [BLANKENHORN, 1999; GUNASEKERA, 2008], it is likely that other types of stress induced the observed upregulation of OxyR-dependent proteins in the intestine of lactose-fed mice. To provide evidence for this suggestion, the *ahpCF* and *dps* gene expression was determined in response to the various substrates occurring in the mouse intestine. These analyses revealed that fermentable carbohydrates, such as glucose, lactose and sorbitol, as well as non-fermentable carbohydrates, such as sucrose [SCHMID, 1982; PIKIS, 2006; LEE, 2010a], induced the *ahpCF* and *dps* promoter activity up to 11-fold under both aerobic and anaerobic growth conditions, whereas dietary proteins did not affect the transcription of these genes. Similar induction levels (up to 14-fold) were also observed after stimulation of cells with the osmolyte NaCl.

These observations are in line with results reported by WEBER et al., who demonstrated the induction of AhpC and Dps by 400 mM NaCl and of Dps by 700 mM sorbitol [WEBER, 2006]. The induction of the *ahpCF* and *dps* expression by carbohydrates and NaCl were absent in *E. coli* mutants lacking the *oxyR* gene. Therefore, it can be concluded that the observed osmolyte-induced expression of *ahpCF* and *dps* in wild type *E. coli* was directly mediated by OxyR. Other transcription factors, such as sigma 24, 38 or 70, binding to the promoter regions of the analysed genes [SAVAGE, 1986;

MOAZED, 1987; ALTUVIA, 1994; KULTZ, 2001], seem to play no or only a minor role. These experiments support the notion that the observed upregulation of AhpF and Dps in intestinal *E. coli* of lactose-fed mice was directly mediated by OxyR.

However, the carbohydrate concentrations required for the *in vitro* induction of *ahpCF* and *dps* were much higher than those detected in the intestinal contents. Assuming an uptake of 5 g diet and 6 ml water per mouse and day [BACHMANOV, 2002], the maximal carbohydrate concentration that is possibly reached in the mouse intestine can be calculated. Accordingly, the lactose concentration in the small intestine would be approximately 130 mM and that of sucrose approximately 270 mM. These theoretical values correspond to those necessary for the induction of the *ahpCF* and *dps* promoters *in vitro* (200 to 400 mM). The differences between the theoretical and the actual concentration of carbohydrates in the mouse gut may result from differences in the intestinal absorption rate, which depends on the varying nutrient availability as influenced by the periods of feeding and starvation [MAGWEDERE, 2008]. As changes at the proteome level are expected to require some time to become effective, proteins that are upregulated in response to the high carbohydrate concentrations in the upper part of the small intestine most likely remain detectable even after absorption of the inducer sucrose by the host.

A common feature of soluble carbohydrates is their osmotic potential. Therefore, it was hypothesised that the carbohydrate-induced expression of OxyR-dependent genes was caused by osmotic stress. This was supported by the positive correlation of the *ahpCF* and *dps* promoter activities with the osmolality of the applied media. Moreover, experiments with aerobically or anaerobically grown *ahpCF* and *oxyR* deletion mutants revealed that OxyR-dependent genes are crucial for *E. coli* to proliferate under conditions of sucrose-mediated high osmolality. Although the deletion of *oxyR* affects genes other than *ahpCF*, the maximal cell densities were similar for both the *ahpCF* and the *oxyR* deletion mutants. The somewhat higher growth inhibition of the *oxyR* mutant manifested itself only by its longer doubling times compared to the *ahpCF* mutant. These results demonstrate that the expression of OxyR-dependent genes, in particular *ahpCF*, enables *E. coli* to better cope with osmotic stress.

Osmotic stress triggers the outflow of water from the bacterial cell resulting in reduced cell turgor and dehydration of the cytoplasm [CSONKA, 1989]. To adapt to such conditions, bacteria are able to take up or synthesise organic osmolytes, such as trehalose, polyols or free amino acids, to enlarge their intracellular solute pool [KEMPF, 1998]. Because the semisynthetic diets fed to the gnotobiotic mice are simple in composition, the intestinal concentration of compatible solutes was expected to be marginal. However, intestinal *E. coli* of mice fed the lactose diet upregulated glutamate dehydrogenase (GdhA), an enzyme catalysing the formation of glutamate, which is required to maintain the steady-state $K^+$ pool after exposure to high osmolality [MCLAGGAN, 1994; YAN, 1996]. This indicates that lactose-induced osmotic stress led to the synthesis of compatible solutes by intestinal *E. coli*.

How can the upregulation of stress-response proteins be explained in view of their physiological role? Are they, similar to the above described expression of GdhA, a supportive mechanism for the protection of *E. coli* against osmotic stress? The OxyR-dependent Fur is a regulatory protein controlling the transcription of genes involved in various processes, such as iron homeostasis, aerobic respiration, chemotaxis, amino acid and purine biosynthesis, carbohydrate metabolism as well as protection against acidic and oxidative stress [CALDERWOOD, 1987; PRODROMOU, 1992; COMPAN, 1993; STOJILJKOVIC, 1994]. Deletion of *fur* in *Desulfovibrio vulgaris* resulted in increased sensitivity to osmotic stress [BENDER, 2007]. Therefore, Fur is

# DISCUSSION

suggested to play a key role in bacterial adaptation to high osmolality but the underlying mechanism is unknown.

The OxyR-regulated *ahpCF* encodes an alkyl hydroperoxide reductase, which reduces alkyl hydroperoxides to their corresponding alcohols [BIEGER, 2001]. Alkyl hydroperoxides, such as lipid hydroperoxides, promote free-radical chain reactions and thereby cause cell membrane and DNA damage [JACOBSON, 1989; LAZIM, 2000]. However, it remains unclear whether, and if yes, in which way osmotic stress causes the formation of intracellular alkyl hydroperoxides.

The OxyR-regulated Dps is involved in the protection of the bacterial cell

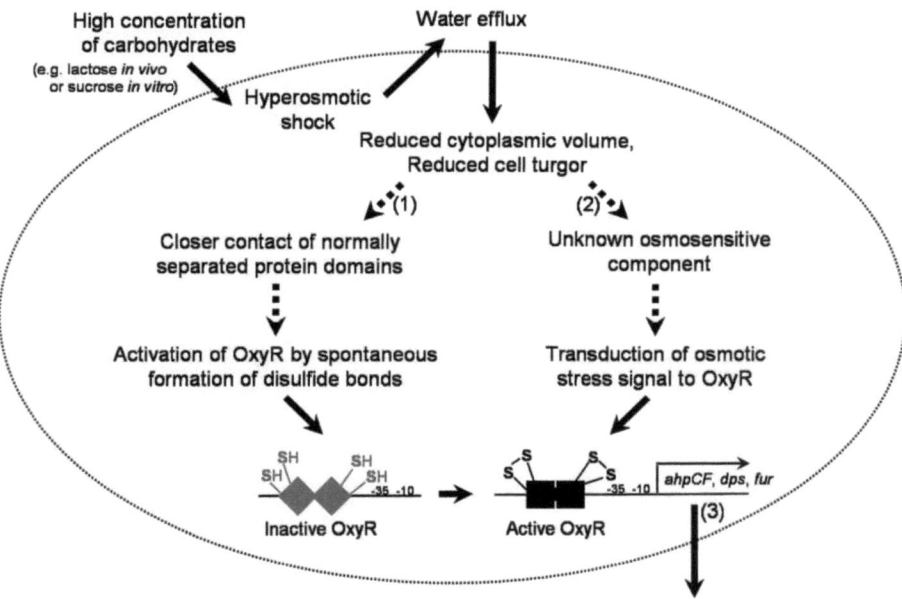

**Figure 26.** Potential mechanisms of OxyR activation by osmotic stress in *E. coli*. Osmotic stress caused by lactose *in vivo* or sucrose *in vitro* leads to reduced cell turgor. This might result in closer contact of intramolecular cysteine sulfhydryl groups of the inactive OxyR (1) or the stress-signal might be transduced to OxyR by an unknown osmosensitive component (2). Both scenarios result in the disulfide bond-mediated activation of OxyR and the transcription of target genes (3) necessary for *E. coli* to better cope with osmotic stress.

against multiple forms of stress, including oxidative stress caused by $H_2O_2$. Dps catalyses direct binding and formation of a DNA-protein crystal [DUKAN, 1996; MARTINEZ, 1997; WOLF, 1999]. KULTZ et al. demonstrated that hyperosmotic stress caused by NaCl leads to DNA double strand breaks in murine kidney cells [KULTZ, 2001]. Such breaks of the DNA phosphodiester backbone usually arise after ionising radiation or $H_2O_2$ treatment [Green, 1996]. That Dps was induced in E. coli after exposure to osmotic stress reveals that DNA protection is also an adaptive mechanism in bacteria.

Several mechanisms underlying the induction of DNA damage by osmotic stress can be envisaged: one possibility is the formation of intracellular free radicals in response to osmotic stress as proposed by ALDSWORTH and DODD. The "suicide through stress" theory of these authors proposes that bacteria generate a burst of reactive oxygen species, such as $H_2O_2$ and $O_2^-$, under various types of stress including heat, osmotic or ethanol stress. This may lead to cell damage and death [ALDSWORTH, 1999; DODD, 2002]. Since these reactive oxygen species were detected only in aerobic organisms with respiratory metabolism (Salmonella enterica serovars, Staphylococcus aureus, Mycobacterium smegmatis, E. coli) but not in strictly fermentative organisms (Streptococcus mutans) the proposed suicide response may be linked to aerobic [DODD, 2007] but not to anaerobic metabolism. The actual study demonstrates the induction of OxyR-dependent genes and their relevance for the growth of E. coli at high osmolality under both aerobic and anaerobic conditions. In addition, incubation of E. coli in the presence of non-fermentable sucrose did not induce the formation of intracellular reactive oxygen species, such as $H_2O_2$. Therefore, other mechanisms of osmotic stress-dependent OxyR activation have to be envisaged (Figure 26).

Such a mechanism might be similar to those operative under oxidative stress conditions, in which OxyR is directly activated via the formation of intermolecular disulfide bonds as described above [ZHENG, 1998]. The hyperosmotic shock and, as a consequence thereof, the rapid water efflux across the cell membrane may cause a closer contact of normally separated intramolecular cysteine sulfhydryl groups as it has been proposed for protein damage after freezing injury [MERYMAN, 1971]. This in turn may lead to the spontaneous formation of disulfide bridges [TAKEDA, 2001] and thereby activate OxyR. It also may be possible that an additional component interacts with OxyR and transduces the osmotic stress signal to OxyR. Nevertheless, these proposed pathways need to be verified by further experiments.

In conclusion, OxyR-dependent stress-response proteins, such as AhpCF, Dps and Fur, are crucial for the adaptation of *E. coli* to osmotic stress conditions. This indicates an overlap of oxidative and osmotic stress responses in *E. coli* and the importance of these responses for the organism's adaptation to carbohydrate-rich host diets.

## 5.4. OxyR-dependent proteins are also required for other commensal bacteria

Since the *oxyR* regulon is a major stress-response system in *E. coli* [STORZ, 1990a; STORZ, 1990b; STORZ, 1990c], it would be of interest to find out whether other microbial species present in the mammalian intestine mount a similar stress response as observed for *E. coli*. OxyR homologues have been identified in various bacterial phyla including Bacteroidetes, Proteobacteria and Firmicutes: the OxyR homologue expressed by *Bacteroides fragilis* controls the expression of *ahpCF*, *dps* and *katB* [ROCHA, 1996; ROCHA, 2000]. In *Pseudomonas aeruginosa*, an OxyR homologue is essential for the synthesis of catalase peroxidase (KatB) and two alkyl hydroperoxide reductases, namely AhpB and AhpCF [OCHSNER, 2000]. In *Klebsiella pneumoniae*, a homologue of *oxyR* was shown to be crucial for the colonisation of the murine gastrointestinal tract [HENNEQUIN, 2009]. In *Bacillus*

# DISCUSSION

*subtilis*, the metalloprotein PerR, a functional analogue of *E. coli*'s OxyR, was identified to control the expression of AhpC, AhpF, KatA and the DNA-binding protein MrgA, which is a Dps homologue [for review see: TRAORE, 2006].

Although these commensal bacteria do not encode the entire *oxyR* regulon present in the *E. coli* genome, the function of OxyR-dependent proteins as well as the activation of OxyR via formation of reversible disulfide bonds between Cys199 and Cys208 in response to $H_2O_2$ is conserved [for review see: CHIANG, 2012]. However, *Clostridium perfringens* and *Bifidobacterium longum* that express genes encoding superoxide dismutases, superoxide reductases, alkyl hydroperoxide reductases and Dps do not encode an OxyR homologous regulatory protein. For these species the underlying regulatory mechanisms are still unknown [JEAN, 2004; XIAO, 2011].

Therefore, the crucial influence of osmotic stress on the gene expression of OxyR-dependent proteins, as presented in the actual study, cannot be directly transferred to other bacterial species. Nevertheless, the presence of OxyR homologues and of OxyR-dependent stress-response proteins, such as Ahp and Dps, in the genome of various commensal bacteria other than *E. coli* strongly indicates that osmotic stress also regulate the OxyR-dependent gene expression in these species. To validate this hypothesis, further experiments using gnotobiotic mice mono-associated with commensal bacteria other than *E. coli* or with a more complex microbial community would be required.

## 5.5. KduI and KduD are required for hexuronate metabolism at high osmolality

Feeding mice the lactose-rich diet not only induced OxyR-dependent stress-response proteins in intestinal *E. coli* but also the so far uncharacterised KduD, while this protein and KduI were downregulated on the casein diet. Homologous enzymes from the plant pathogen *Erwinia chrysanthemi* are known to catalyse the degradation of pectin. Polygalacturonate and galacturonate resulting from pectin catabolism were shown to induce *kdul* gene expression in *E. chrysanthemi* [CONDEMINE, 1991]. However, *E. coli* does not possess enzymes involved in the utilisation of pectin as well as poly- and oligogalacturonates [RODIONOV, 2004]. Therefore, the role of KduI and KduD in the *E. coli* metabolism is obscure and the reasons for the induction of KduD in intestinal *E. coli* of lactose-fed mice are unknown.

To identify the nature of the upregulation of KduD in intestinal *E. coli*, the expression of the corresponding gene in response to the various substrates occurring in the mouse intestine was determined. This analysis revealed that only galacturonate and glucuronate, which are not present in the mouse diets, induced the *kduD* promoter activity up to 3-fold under aerobic and up to 20-fold under anaerobic conditions. These hexuronates also increased the promoter activity of *kduI*. However, as observed for *kduD*, higher induction levels were detected under anaerobic compared to aerobic growth conditions. This suggests that the *kduI* and *kduD* gene expression was facilitated in an anaerobic environment. Transcription factors involved in gene regulation under anaerobic conditions, such as the transcriptional regulator FNR (fumarate and nitrate reductase regulator), the ArcAB two-component system [SALMON, 2005; KANG, 2005; MALPICA, 2006] or sigma factors (sigma 24, 38 or 70) [SAVAGE, 1986; MOAZED, 1987; ALTUVIA, 1994; KULTZ, 2001] may bind to the *kduI* and *kduD* promoters and possibly enhance gene expression. However, more detailed analyses are required to elucidate the regulatory systems involved in hexuronate-induced *kduI* and *kduD* gene expression under anaerobic conditions.

In fact, hexuronate-induced *kduI* gene expression was 2 to 4-fold higher than that of *kduD* under aerobic and anaerobic conditions. In contrast to the native KduD, which contain only one polypeptide chain, the native KduI forms a hexamer in solution [DUNTEN, 1998]. Therefore, more KduI-subunits are required to form a functional enzyme. This may explain the higher expression level observed for *kduI*. Nevertheless, the underlying molecular mechanism how galacturonate and glucuronate regulate *kduI* and *kduD* gene expression is currently unknown.

Experiments with cell-free extracts of *E. coli* overexpressing KduI, KduD or both revealed a role of KduI in the degradation of galacturonate and glucuronate. However, how KduI facilitates the conversion of these hexuronates is unclear. KduI may activate as yet unknown alternative hexuronate degrading pathways, thereby facilitating the conversion of galacturonate and glucuronate. However, the gene expression of common hexuronate degrading enzymes, such as UxaABC and UxuAB, is only induced in the presence of the corresponding substrate (galacturonate, glucuronate, fructuronate etc.) [PORTALIER, 1972; ROBERT-BAUDOUY, 1973; PORTALIER, 1980]. Since these substrates were not present in the medium used for cultivation of the cells overexpressing KduI, it is unlikely that other alternative hexuronate degrading enzymes are expressed constitutively. Therefore, a catabolic role of KduI in the utilisation of galacturonate and glucuronate is the most probable explanation.

In *E. coli*'s regular hexuronate degrading pathway, galacturonate and glucuronate enter the bacterial cell via the aldohexuronate transporter (ExuT) [MATA-GILSINGER, 1983] and subsequently undergo isomerisation to tagaturonate or fructuronate by uronate isomerase (UxaC). In the next step, altronate oxidoreductase (UxaB) or mannonate oxidoreductase (UxuB) catalyse the reduction of tagaturonate or fructuronate to altronate or mannonate with the simultaneous oxidation of NADH. Altronate and mannonate are converted by altronate dehydratase (UxaA) or mannonate dehydratase (UxuA) to 2-oxo-3-deoxygluconate [Ritzenthaler, 1981; MATA-GILSINGER, 1983], which enters glycolysis (Figure 27B).

Do *E. coli*'s KduI and KduD have the capacity to catalyse the conversion of galacturonate and glucuronate, as well? Based on 75% sequence identity of

# DISCUSSION

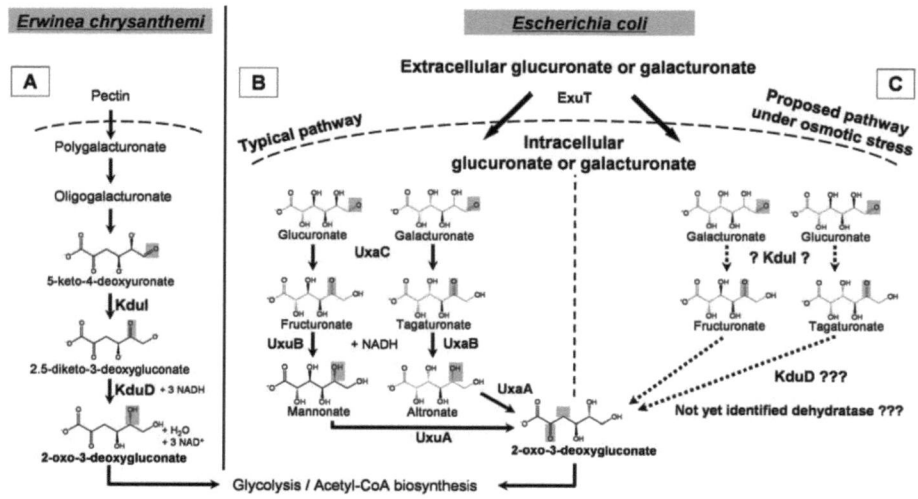

Figure 27. Role of KduI and KduD in *E. chrysanthemi* and *E. coli*. (A) Reactions catalysed by 5-keto 4-deoxyuronate isomerase (**KduI**) and 2-deoxy-D-gluconate 3-dehydrogenase (**KduD**) in *E. chrysanthemi*. (B) Typical pathway of hexuronate conversion in *E. coli*: galacturonate and glucuronate enter the bacterial cell via the aldohexuronate transporter (**ExuT**) and undergo isomerisation to tagaturonate or fructuronate by the uronate isomerase (**UxaC**). The NADH-dependent reduction to altronate or mannonate is catalysed by the altronate oxidoreductase (**UxaB**) or mannonate oxidoreductase (**UxuB**). These intermediates are further converted by the altronate dehydratase (**UxaA**) or mannonate dehydratase (**UxuA**) to 2-oxo-3-deoxygluconate, which subsequently enter glycolysis. (C) The results of the actual work indicate an alternative hexuronate degrading pathway in *E. coli*: under osmotic stress, KduI compensates for reduced levels of the typical hexuronate degrading enzyme UxaC, which gene expression is repressed under these conditions. Based on sequence similarities to the KduI of *E. chrysanthemi* and on the closely related structure of galacturonate and glucuronate to that of 5-keto 4-deoxyuronate, *E. coli*'s KduI was hypothesised to catalyse the isomerisation of galacturonate and glucuronate to tagaturonate and fructuronate. Although the data also demonstrated the hexuronate-driven induction of KduD, a function of this enzyme in hexuronate breakdown remains unclear. Black arrows indicate experimental evidence, dashed arrow indicate hypothesised reactions.

the *E. coli* KduI and the 5-keto 4-deoxyuronate isomerase of *E. chrysanthemi* (Figure 27A), the *E. coli* KduI is predicted to catalyse the same reaction [KESELER, 2011; KESELER, 2012]. Based on this and on the closely related structure of galacturonate and glucuronate to that of 5-keto-4-deoxyuronate, *E. coli*'s KduI may catalyse the isomerisation of galacturonate and glucuronate to tagaturonate and fructuronate. If this hypothesis is true, KduI may compensate for the function of UxaC, whose gene expression was repressed by $H_2O_2$-induced oxidative stress and osmotic stress by up to 60%

in an OxyR-dependent manner (Figure 27C).

Based on an *E. chrysanthemi* enzyme that is homologous to KduD and catalyses the NADH-dependent reduction of 2,5-dioxo-3-deoxygluconate to 2-oxo-3-deoxygluconate (Figure 27A) [CONDEMINE, 1991], *E. coli*'s KduD is predicted to be a 2-deoxy-D-gluconate 3-dehydrogenase [KESELER, 2011; KESELER, 2012]. Although KduD was upregulated in intestinal *E. coli* of mice fed the lactose-rich diet and *kduD* gene expression was induced by galacturonate and glucuronate, this enzyme did not facilitate the conversion of these hexuronates in cell-free extracts and, hence, do not compensate for the function of UxaC. However, a role of this protein in further steps of hexuronate degradation cannot be excluded because only galacturonate and glucuronate but not tagaturonate and fructuronate were used as substrates in the experiments. Nevertheless, in the regular hexuronate degrading pathway, the reduction of tagaturonate and fructuronate to altronate and mannonate is catalysed by different enzymes, namely UxaB and UxuB (Figure 27B, C). It is unlikely that KduD substitutes for the function of both UxaB and UxuB, although their corresponding genes, as mentioned above for *uxaC*, were repressed by osmotic stress. Therefore, additional experiments and more detailed characterisation of KduI and KduD would be required to verify the isomerase activity of KduI on the one hand and to get insight into the activity of KduD on the other hand. However, it is presently not known, whether an alternative enzyme may compensate for the function of UxaA or UxuA and may convert altronate or mannonate to 2-oxo-3-deoxygluconate.

In contrast to *uxaC*, *uxaB* and *uxuAB*, the hexuronate-driven *kduI* and *kduD* gene expression was OxyR-independent and was not repressed by osmotic stress. Interestingly, in the presence of galacturonate, osmotic stress led to approximately 2-fold higher *kduI* and *kduD* gene expression, while on glucuronate no differences were observed (both in comparison to incubation on galacturonate or glucuronate alone). The mechanism underlying this observation is so far unknown. Two explanations are possible: (i) other transcription factors involved in stress response processes, such as sigma 24, 38 or 70 [SAVAGE, 1986; MOAZED, 1987; ALTUVIA, 1994; KULTZ, 2001], may bind to the promoter regions of *kduI* and *kduD* and facilitate their transcription. (ii) An unknown osmosensitive stress sensor may transduce the osmotic stress signal and enhance the *kduI* and *kduD* transcription at high osmolality. Although the structure of galacturonate and glucuronate is closely related, the probable stress-response sensor may be more susceptible to galacturonate than to glucuronate.

The essential physiological role of KduI and KduD in hexuronate metabolism under osmotic stress conditions, in which the regular hexuronate degrading enzymes are repressed, was supported by growth experiments with mutants lacking the *kduID* genes (Figure 28). Therefore, the question arose whether galacturonate and glucuronate were present in the intestines of the gnotobiotic mice. Even though hexuronates were not present in the mouse diets (see section 3.1.1.) they appear as sugar substituents of glycoproteins and glycolipids in the mammalian mucus layer [ALLEN, 1984; PEEKHAUS, 1998] and are therefore expected to occur in intestinal contents of mice. Although the *E. coli* genome encode approximately 40 glycoside hydrolases [HENRISSAT, 1997], this organism cannot degrade complex mucus polysaccharides [HOSKINS, 1985] and is limited to growth on simple mono- and oligosaccharides released from mucosal glycoproteins [CHANG, 2004]. Despite the detection of slightly higher hexuronate concentrations in the intestines of lactose-fed mice, it is presently not known, whether these hexuronates originate from lactose and, if yes, which pathways are used for their formation. The transient formation of cytoplasmic hexuronates in *E. coli* grown on lactose but not on glucose supports the hypothesis that the detected intracellular hexuronates originate from lactose. However, to clearly identify possible sources or precursors of these detected intracellular

# DISCUSSION

hexuronates, experiments using isotopically labelled lactose would be required. Presently, it can only be speculated how these lactose-derived hexuronates could be generated from lactose. The most likely explanation for lactose-derived hexuronates is the generation of hexuronate precursors during lactose catabolism. In *E. coli*, lactose cleavage by ß-galactosidase results in the formation of galactose and glucose [COHN, 1989; BECKWITH, 1967]. While glucose directly enters glycolysis, galactose is usually converted by Leloir pathway enzymes [FREY, 1996]. Intracellular hexuronates were only detected in exponentially growing *E. coli*. Therefore, the high

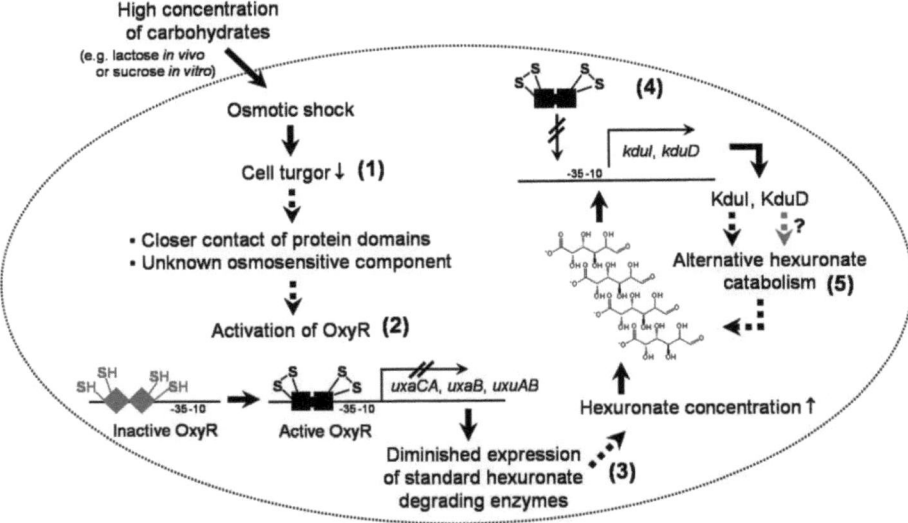

Figure 28. **Potential mechanism of how osmotic stress may influence hexuronate degrading enzymes.** High carbohydrate concentrations caused by lactose *in vivo* or sucrose *in vitro* induces osmotic stress (1), which is proposed to activate the OxyR transcriptional regulator by a closer contact of normally separated protein domains or an unknown osmosensitive transducing component. The genes of the regular hexuronate degrading enzymes (uronate isomerase, UxaC; altronate oxidoreductase, UxaB; mannonate oxidoreductase, UxuB; altronate dehydratase, UxaA; mannonate dehydratase, UxuA) are repressed in an OxyR-dependent manner (2), which may diminish the hexuronate utilisation in the presence of osmotic stress (3). In contrast, expression of *kduI* and *kduD*, which is inducible by galacturonate and glucuronate, is neither OxyR-dependent nor repressed by osmotic stress (4). The 5-keto 4-deoxyuronate isomerase (KduI) is hypothesised to catalyse or facilitate the conversion of galacturonate and glucuronate and therefore compensate for the activity of UxaC under conditions of high osmolality. However, the role of the 2-deoxy-D-gluconate 3-dehydrogenase (KduD) in hexuronate conversion is presently unclear (5). Black arrows indicate experimental evidence, dashed arrow indicate hypothesised relations.

metabolic rate characterising this growth phase may result in the accumulation of either intracellular UDP-glucose or galactose as described for lactic acid bacteria [KLEEREBEZEM, 1999; HUGENHOLTZ, 1999]. Since the uronate dehydrogenase, which was used for the hexuronate determination, cannot distinguish between galacturonate and glucuronate [MOON, 2009], the source of the detected intracellular hexuronates could not be clearly identified. Two explanations are possible: 1. Intracellular glucuronate may stem from UDP-glucuronate, which in turn is generated from UDP-glucose as catalysed by UDP-glucose 6-dehydrogenase [STEVENSON, 1996]. 2. Intracellular galacturonate may stem from oxidation of accumulated galactose. However, such a reaction has not yet been demonstrated in *E. coli*.

The detected disappearance of intracellular hexuronates during transition of *E. coli* from exponential to stationary growth phase suggests that they serve as substrates. In lactose-fed mice, lactose is readily available during the feeding periods and enables intestinal *E. coli* to grow exponentially. This would induce the constant formation of cytoplasmic hexuronates, which in turn may provoke the expression of *kduI* and *kduD*, whose corresponding enzymes are suggested to be involved in hexuronate conversion.

Is cytoplasmic galacturonate or glucuronate responsible for the elevated intestinal hexuronate concentrations in lactose-fed mice? In *E. coli*, intracellular glucuronate can be exported via the gluconate transporter (GntT) or facilitated diffusion mechanisms [RICHEY, 1972; ABENDANO, 1973; LAGARDE, 1975; LANCASTER, 1977; KORNBERG, 2000]. However, a release of hexuronates would not benefit *E. coli* because they are potential energy sources. Furthermore, the cytoplasmic hexuronate concentrations of *E. coli* grown on lactose-containing medium were similar but not higher than those in the intestinal contents of lactose-fed mice. Therefore, export of hexuronates is unlikely to happen and cannot be responsible for the observed intestinal hexuronate concentrations.

Taken together, these experiments suggest that *kduI* and *kduD* are upregulated in intestinal *E. coli* of lactose-fed mice in response to increased galacturonate and glucuronate concentrations, whose conversion was facilitated by KduI. These results indicate a novel function of KduI in *E. coli* and demonstrate the crucial influence of diet-induced osmotic stress on the gene expression of *E. coli*'s regular hexuronate degrading enzyme UxaC, whose function are possibly compensated by KduI.

## 5.6. Concluding remarks

The aim of this study was to identify the impact of dietary factors on the protein expression of intestinal *E. coli*. The results demonstrate the adaptation of intestinal *E. coli* to the dietary conditions in the mouse intestine, as obvious from the induction of enzymes belonging to the Leloir pathway in response to feeding mice the lactose-rich diet. Novel functions of enzymes belonging to stress-response and carbohydrate catabolic processes could be identified. On the one hand, genes typically upregulated in response to oxidative stress (*ahpCF*, *dps*) were shown to be sensitive to carbohydrate-induced osmotic stress as well. On the other hand, the gene expression of common hexuronate degrading enzymes (*uxaCA*, *uxaB*, *uxuAB*) was repressed under such conditions, while an enzyme with as yet unknown function (KduI) was suggested to compensate for reduced levels of UxaC. These results indicate that osmotic stress not only acts as an enhancer but also as a repressor of the expression of OxyR-dependent genes and suggest an overlap of *E. coli*'s oxidative and osmotic stress responses, whose role in the organism's adaptation to a carbohydrate-rich host diet has been demonstrated in this thesis.

JONES and FABICH postulated that *E. coli* are dependent not only on anaerobic but also on microaerobic respiration to successfully colonise the mammalian intestine, in which the oxygen partial pressure is low, but not completely anaerobic: enzymes typically required for aerobic respiration, such as ATP synthase, cytochrome *bd* oxidase and cytochrome *bo$_3$* oxidase, has been demonstrated to be essential for *E. coli* to colonise the mouse intestine [JONES, 2007; FABICH, 2008]. However, because of the crucial influence of osmotic stress on the expression of OxyR-dependent oxidative stress-response proteins (AhpF, Dps), as presented in the actual study, the role of reactive oxygen species generated during aerobic respiration might have been overestimated: not only the oxygen partial pressure but also osmotically active nutritional compounds, such as dietary carbohydrates, may stimulate the expression of genes typically required for the adaptation to an aerobic environment and, hence, improve the metabolic flexibility of intestinal *E. coli*.

Since OxyR homologues and OxyR-dependent genes such as *ahpCF*, *dps* and *fur* have also been identified in commensal bacteria other than *E. coli* [ROCHA, 2000; OCHSNER, 2000; JEAN, 2004; HENNEQUIN, 2009; XIAO, 2011], the ability to express oxidative stress-response proteins in response to osmotic stress would enhance the respiratory flexibility and facilitate colonisation of the intestinal environment by these species [PEEKHAUS, 1998; CHANG, 2004; JONES, 2007; VOGEL-SCHEEL, 2010]. Therefore, it may be speculated that the observed alteration of the intestinal microbiota composition in response to changing nutritional factors [FAVIER, 2002; LEY, 2008; Turnbaugh, 2009b; FLEISSNER, 2010; FAITH, 2011; WU, 2011; PATRONE, 2012; YATSUNENKO, 2012] is associated with the ability to adapt to diet-induced osmotic stress. Nevertheless, to verify this assumption, further experiments using gnotobiotic mice associated with commensal intestinal bacteria other than *E. coli* or with a simplified community of representative intestinal species are required.

## APPENDIX I: REFERENCES

1  Abendano JJ and Kepes A (1973). **Sensitization of D-glucuronic acid transport system of *E. coli* to protein group reagents in presence of substrate or absence of energy source.** *Biochem Biophys Res Commun* 54: 1342-1346.

2  Adlerberth I, Cerquetti M, Poilane I, Wold A and Collignon A (2000). **Mechanisms of Colonisation and Colonisation Resistance of the Digestive Tract.** *Microb Ecol Health Dis* 12: 223-239.

3  Aldsworth TG, Sharman RL and Dodd CE (1999). **Bacterial suicide through stress.** *Cell Mol Life Sci* 56: 378-383.

4  Allen A (1984). **The structure and function of gastrointestinal mucus.** In: Boedeker IEC (ed) *Attachment of organisms to the gut mucosa* vol II. CRC Press, Boca Raton, FL: pp 3-11.

5  Alpert C, Engst W, Guehler A, Oelschlaeger T and Blaut M (2005). **Bacterial response to eukaryotic cells. Analysis of differentially expressed proteins using nano liquid chromatography-electrospray ionization tandem mass spectrometry.** *J Chromatogr A* 1082: 25-32.

6  Alpert C, Scheel J, Engst W, Loh G and Blaut M (2009). **Adaptation of protein expression by *Escherichia coli* in the gastrointestinal tract of gnotobiotic mice.** *Environ Microbiol* 11: 751-761.

7  Altuvia S, Almiron M, Huisman G, Kolter R and Storz G (1994). **The *dps* promoter is activated by OxyR during growth and by IHF and sigma S in stationary phase.** *Mol Microbiol* 13: 265-272.

8  Araneo BA, Cebra JJ, Beuth J, Fuller R, Heidt PJ, Midvedt T, Nord CE et al. (1996). **Problems and priorities for controlling opportunistic pathogens with new antimicrobial strategies; an overview of current literature.** *Zentralbl Bakteriol* 283: 431-465.

9  Baba T, Ara T, Hasegawa M, Takai Y, Okumura Y, Baba M, Datsenko KA et al. (2006). **Construction of *Escherichia coli* K-12 in-frame,**

single-gene knockout mutants: the Keio collection. *Mol Syst Biol* 2: 2006 0008.

10 Bachmanov AA, Reed DR, Beauchamp GK and Tordoff MG (2002). **Food intake, water intake, and drinking spout side preference of 28 mouse strains.** *Behav Genet* 32: 435-443.

11 Beaulieu JF and Quaroni A (1991). **Clonal analysis of sucrase-isomaltase expression in the human colon adenocarcinoma Caco-2 cells.** *Biochem J* 280 ( Pt 3): 599-608.

12 Beckwith JR (1967). **Regulation of the *lac* operon. Recent studies on the regulation of lactose metabolism in *Escherichia coli* support the operon model.** *Science* 156: 597-604.

13 Bender KS, Yen HC, Hemme CL, Yang Z, He Z, He Q, Zhou J et al. (2007). **Analysis of a ferric uptake regulator (Fur) mutant of *Desulfovibrio vulgaris* Hildenborough.** *Appl Environ Microbiol* 73: 5389-5400.

14 Bengmark S (1998). **Ecological control of the gastrointestinal tract. The role of probiotic flora.** *Gut* 42: 2-7.

15 Bieger B and Essen LO (2001). **Crystal structure of the catalytic core component of the alkylhydroperoxide reductase AhpF from *Escherichia coli*.** *J Mol Biol* 307: 1-8.

16 Blankenhorn D, Phillips J and Slonczewski JL (1999). **Acid- and base-induced proteins during aerobic and anaerobic growth of *Escherichia coli* revealed by two-dimensional gel electrophoresis.** *J Bacteriol* 181: 2209-2216.

17 Blattner FR, Plunkett G, 3rd, Bloch CA, Perna NT, Burland V, Riley M, Collado-Vides J et al. (1997). **The complete genome sequence of *Escherichia coli* K-12.** *Science* 277: 1453-1462.

18 Bradford MM (1976). **A rapid and sensitive method for the quantitation of microgram quantities of protein utilizing the principle of protein-dye binding.** *Anal Biochem* 72: 248-254.

19 Brown G, Singer A, Proudfoot M, Skarina T, Kim Y, Chang C, Dementieva I et al. (2008). **Functional and structural**

characterization of four glutaminases from *Escherichia coli* and *Bacillus subtilis*. *Biochemistry* 47: 5724-5735.

20  Bry L, Falk PG, Midtvedt T and Gordon JI (1996). **A model of host-microbial interactions in an open mammalian ecosystem**. *Science* 273: 1380-1383.

21  Butterton JR, Ryan ET, Shahin RA and Calderwood SB (1996). **Development of a germfree mouse model of *Vibrio cholerae* infection**. *Infect Immun* 64: 4373-4377.

22  Cabiscol E, Tamarit J and Ros J (2000). **Oxidative stress in bacteria and protein damage by reactive oxygen species**. *Int Microbiol* 3: 3-8.

23  Calderwood SB and Mekalanos JJ (1987). **Iron regulation of Shiga-like toxin expression in *Escherichia coli* is mediated by the *fur* locus**. *J Bacteriol* 169: 4759-4764.

24  Chang DE, Smalley DJ, Tucker DL, Leatham MP, Norris WE, Stevenson SJ, Anderson AB et al. (2004). **Carbon nutrition of *Escherichia coli* in the mouse intestine**. *Proc Natl Acad Sci U S A* 101: 7427-7432.

25  Chiang SM and Schellhorn HE (2012). **Regulators of oxidative stress response genes in *Escherichia coli* and their functional conservation in bacteria**. *Arch Biochem Biophys* 525: 161-169.

26  Christman MF, Storz G and Ames BN (1989). **OxyR, a positive regulator of hydrogen peroxide-inducible genes in *Escherichia coli* and *Salmonella typhimurium*, is homologous to a family of bacterial regulatory proteins**. *Proc Natl Acad Sci U S A* 86: 3484-3488.

27  Clapper WE and Meade GH (1963). **Normal Flora of the Nose, Throat, and Lower Intestine of Dogs**. *J Bacteriol* 85: 643-648.

28  Cohn M (1989). **The way it was: a commentary on Studies on the Induced Synthesis of beta-galactosidase in *Escherichia coli*: the Kinetics and Mechanism of Sulfur Incorporation**. *Biochim Biophys Acta* 1000: 109-112.

29  Compan I and Touati D (1993). **Interaction of six global transcription regulators in expression of manganese superoxide dismutase in *Escherichia coli* K-12**. *J Bacteriol* 175: 1687-1696.

30  Comstock LE and Coyne MJ (2003). **Bacteroides thetaiotaomicron: a dynamic, niche-adapted human symbiont**. *Bioessays* 25: 926-929.

31  Condemine G, Hugouvieux-Cotte-Pattat N and Robert-Baudouy J (1986). **Isolation of *Erwinia chrysanthemi kduD* mutants altered in pectin degradation**. *J Bacteriol* 165: 937-941.

32  Condemine G and Robert-Baudouy J (1991). **Analysis of an *Erwinia chrysanthemi* gene cluster involved in pectin degradation**. *Mol Microbiol* 5: 2191-2202.

33  Crowther RL and Georgiadis MM (2005). **The crystal structure of 5-keto-4-deoxyuronate isomerase from *Escherichia coli***. *Proteins* 61: 680-684.

34  Csonka LN (1989). **Physiological and genetic responses of bacteria to osmotic stress**. *Microbiol Rev* 53: 121-147.

35  Datsenko KA and Wanner BL (2000). **One-step inactivation of chromosomal genes in *Escherichia coli* K-12 using PCR products**. *Proc Natl Acad Sci U S A* 97: 6640-6645.

36  de Sousa RR, Queiroz KC, Souza AC, Gurgueira SA, Augusto AC, Miranda MA, Peppelenbosch MP et al. (2007). **Phosphoprotein levels, MAPK activities and NFkappaB expression are affected by fisetin**. *J Enzyme Inhib Med Chem* 22: 439-444.

37  Denou E, Rezzonico E, Panoff JM, Arigoni F and Brussow H (2009). **A Mesocosm of *Lactobacillus johnsonii*, *Bifidobacterium longum*, and *Escherichia coli* in the mouse gut**. *DNA Cell Biol* 28: 413-422.

38  Dodd CE and Aldsworth TG (2002). **The importance of RpoS in the survival of bacteria through food processing**. *Int J Food Microbiol* 74: 189-194.

39  Dodd CE, Richards PJ and Aldsworth TG (2007). **Suicide through stress: a bacterial response to sub-lethal injury in the food environment**. *Int J Food Microbiol* 120: 46-50.

40  Dower WJ, Miller JF and Ragsdale CW (1988). **High efficiency transformation of E. coli by high voltage electroporation.** *Nucleic Acids Res* 16: 6127-6145.

41  Dukan S and Touati D (1996). **Hypochlorous acid stress in Escherichia coli: resistance, DNA damage, and comparison with hydrogen peroxide stress.** *J Bacteriol* 178: 6145-6150.

42  Dunten P, Jaffe H and Aksamit RR (1998). **Crystallization of 5-keto-4-deoxyuronate isomerase from Escherichia coli.** *Acta Crystallogr D Biol Crystallogr* 54: 678-680.

43  Duobos R, Schaedler RW and Costello R (1963). **Composition, Alteration, and Effects of the Intestinal Flora.** *Fed Proc* 22: 1322-1329.

44  Escherich T (1988). **The intestinal bacteria of the neonate and breast-fed infant. 1884.** *Rev Infect Dis* 10: 1220-1225.

45  Espey MG (2013). **Role of oxygen gradients in shaping redox relationships between the human intestine and its microbiota.** *Free Radic.Biol.Med.* 55: 130-140.

46  Fabich AJ, Jones SA, Chowdhury FZ, Cernosek A, Anderson A, Smalley D, McHargue JW et al. (2008). **Comparison of carbon nutrition for pathogenic and commensal Escherichia coli strains in the mouse intestine.** *Infect Immun* 76: 1143-1152.

47  Faith JJ, McNulty NP, Rey FE and Gordon JI (2011). **Predicting a human gut microbiota's response to diet in gnotobiotic mice.** *Science* 333: 101-104.

48  Falk PG, Hooper LV, Midtvedt T and Gordon JI (1998). **Creating and maintaining the gastrointestinal ecosystem: what we know and need to know from gnotobiology.** *Microbiol Mol Biol Rev* 62: 1157-1170.

49  Favier CF, Vaughan EE, De Vos WM and Akkermans AD (2002). **Molecular monitoring of succession of bacterial communities in human neonates.** *Appl Environ Microbiol* 68: 219-226.

50    Ferraris RP (2001). **Dietary and developmental regulation of intestinal sugar transport.** *Biochem J* 360: 265-276.

51    Fleissner CK, Huebel N, Abd El-Bary MM, Loh G, Klaus S and Blaut M (2010). **Absence of intestinal microbiota does not protect mice from diet-induced obesity.** *Br J Nutr* 104: 919-929.

52    Franks AH, Harmsen HJ, Raangs GC, Jansen GJ, Schut F and Welling GW (1998). **Variations of bacterial populations in human feces measured by fluorescent in situ hybridization with group-specific 16S rRNA-targeted oligonucleotide probes.** *Appl Environ Microbiol* 64: 3336-3345.

53    Freter R, Brickner H, Botney M, Cleven D and Aranki A (1983). **Mechanisms that control bacterial populations in continuous-flow culture models of mouse large intestinal flora.** *Infect Immun* 39: 676-685.

54    Freundlieb S and Boos W (1986). **Alpha-amylase of *Escherichia coli*, mapping and cloning of the structural gene, *malS*, and identification of its product as a periplasmic protein.** *J Biol Chem* 261: 2946-2953.

55    Frey PA (1996). **The Leloir pathway: a mechanistic imperative for three enzymes to change the stereochemical configuration of a single carbon in galactose.** *Faseb J* 10: 461-470.

56    Froger A and Hall JE (2007). **Transformation of plasmid DNA into *E. coli* using the heat shock method.** *J Vis Exp* 253.

57    Gibson GR (2004). **Fibre and effects on probiotics (the prebiotic concept).** *Clinical Nutrition Supplements* 1: 25–31.

58    Gibson GR, Beatty ER, Wang X and Cummings JH (1995). **Selective stimulation of bifidobacteria in the human colon by oligofructose and inulin.** *Gastroenterology* 108: 975-982.

59    Gill SR, Pop M, Deboy RT, Eckburg PB, Turnbaugh PJ, Samuel BS, Gordon JI et al. (2006). **Metagenomic analysis of the human distal gut microbiome.** *Science* 312: 1355-1359.

## REFERENCES

60  Görlach A and Kietzmann T (2007). **Superoxide and derived reactive oxygen species in the regulation of hypoxia-inducible factors.** *Methods Enzymol* 435: 421-446.

61  Green MH, Lowe JE, Delaney CA and Green IC (1996). **Comet assay to detect nitric oxide-dependent DNA damage in mammalian cells.** *Methods Enzymol* 269: 243-266.

62  Gunasekera TS, Csonka LN and Paliy O (2008). **Genome-wide transcriptional responses of *Escherichia coli* K-12 to continuous osmotic and heat stresses.** *J Bacteriol* 190: 3712-3720.

63  He G, Shankar RA, Chzhan M, Samouilov A, Kuppusamy P and Zweier JL (1999). **Noninvasive measurement of anatomic structure and intraluminal oxygenation in the gastrointestinal tract of living mice with spatial and spectral EPR imaging.** *Proc Natl Acad Sci U S A* 96: 4586-4591.

64  Henderson LM and Chappell JB (1993). **Dihydrorhodamine 123: a fluorescent probe for superoxide generation?** *Eur J Biochem* 217: 973-980.

65  Hennequin C and Forestier C (2009). ***oxyR*, a LysR-type regulator involved in *Klebsiella pneumoniae* mucosal and abiotic colonization.** *Infect Immun* 77: 5449-5457.

66  Henrissat B and Davies G (1997). **Structural and sequence-based classification of glycoside hydrolases.** *Curr Opin Struct Biol* 7: 637-644.

67  Holden HM, Rayment I and Thoden JB (2003). **Structure and function of enzymes of the Leloir pathway for galactose metabolism.** *J Biol Chem* 278: 43885-43888.

68  Hooper LV, Littman DR and Macpherson AJ (2012). **Interactions between the microbiota and the immune system.** *Science* 336: 1268-1273.

69  Hooper LV, Midtvedt T and Gordon JI (2002). **How host-microbial interactions shape the nutrient environment of the mammalian intestine.** *Annu Rev Nutr* 22: 283-307.

70  Hooper LV, Xu J, Falk PG, Midtvedt T and Gordon JI (1999). **A molecular sensor that allows a gut commensal to control its nutrient foundation in a competitive ecosystem**. *Proc Natl Acad Sci U S A* 96: 9833-9838.

71  Hoskins LC, Agustines M, McKee WB, Boulding ET, Kriaris M and Niedermeyer G (1985). **Mucin degradation in human colon ecosystems. Isolation and properties of fecal strains that degrade ABH blood group antigens and oligosaccharides from mucin glycoproteins**. *J Clin Invest* 75: 944-953.

72  Hu W, Tedesco S, McDonagh B, Barcena JA, Keane C and Sheehan D (2010). **Selection of thiol- and disulfide-containing proteins of *Escherichia coli* on activated thiol-Sepharose**. *Anal Biochem* 398: 245-253.

73  Hudault S, Guignot J and Servin AL (2001). ***Escherichia coli* strains colonising the gastrointestinal tract protect germfree mice against *Salmonella typhimurium* infection**. *Gut* 49: 47-55.

74  Hugenholtz J and Kleerebezem M (1999). **Metabolic engineering of lactic acid bacteria: overview of the approaches and results of pathway rerouting involved in food fermentations**. *Curr Opin Biotechnol* 10: 492-497.

75  Hwang MN and Ederer GM (1975). **Rapid hippurate hydrolysis method for presumptive identification of group B *streptococci***. *J Clin Microbiol* 1: 114-115.

76  Ingledew WJ and Poole RK (1984). **The respiratory chains of *Escherichia coli***. *Microbiol Rev* 48: 222-271.

77  Jacobson FS, Morgan RW, Christman MF and Ames BN (1989). **An alkyl hydroperoxide reductase from *Salmonella typhimurium* involved in the defense of DNA against oxidative damage. Purification and properties**. *J Biol Chem* 264: 1488-1496.

78  Jean D, Briolat V and Reysset G (2004). **Oxidative stress response in *Clostridium perfringens***. *Microbiology* 150: 1649-1659.

79  Jones SA, Chowdhury FZ, Fabich AJ, Anderson A, Schreiner DM, House AL, Autieri SM et al. (2007). **Respiration of *Escherichia coli* in the mouse intestine.** *Infect Immun* 75: 4891-4899.

80  Kageyama A, Benno Y and Nakase T (1999). **Phylogenetic evidence for the transfer of *Eubacterium lentum* to the genus Eggerthella as *Eggerthella lenta* gen. nov., comb. nov.** *Int J Syst Bacteriol* 49 Pt 4: 1725-1732.

81  Kamiya S, Taguchi H, Yamaguchi H, Osaki T, Takahashi M and Nakamura S (1997). **Bacterioprophylaxis using *Clostridium butyricum* for lethal caecitis by *Clostridium difficile* in gnotobiotic mice.** *Rev. Med. Microbiol* 8(Suppl. 1): S57-S59.

82  Kamlage B, Hartmann L, Gruhl B and Blaut M (1999). **Intestinal microorganisms do not supply associated gnotobiotic rats with conjugated linoleic acid.** *J Nutr* 129: 2212-2217.

83  Kang Y, Weber KD, Qiu Y, Kiley PJ and Blattner FR (2005). **Genome-wide expression analysis indicates that FNR of *Escherichia coli* K-12 regulates a large number of genes of unknown function.** *J Bacteriol* 187: 1135-1160.

84  Kaur K, Mahmood S and Mahmood A (2006). **Susceptibility of lactase to luminal proteases in developing rat intestine.** *Indian J Gastroenterol* 25: 179-181.

85  Kawano M, Abuki R, Igarashi K and Kakinuma Y (2000). **Evidence for Na(+) influx via the NtpJ protein of the KtrII K(+) uptake system in *Enterococcus hirae*.** *J Bacteriol* 182: 2507-2512.

86  Kempf B and Bremer E (1998). **Uptake and synthesis of compatible solutes as microbial stress responses to high-osmolality environments.** *Arch Microbiol* 170: 319-330.

87  Keseler IM, Collado-Vides J, Santos-Zavaleta A, Peralta-Gil M, Gama-Castro S, Muniz-Rascado L, Bonavides-Martinez C et al. (2011). **EcoCyc: a comprehensive database of *Escherichia coli* biology.** *Nucleic Acids Res* 39: D583-590.

88  Keseler IM, Mackie A, Peralta-Gil M, Santos-Zavaleta A, Gama-Castro S, Bonavides-Martinez C, Fulcher C et al. (2012). **EcoCyc: fusing**

model organism databases with systems biology. *Nucleic Acids Res.*

89  Kim YS and Ho SB (2010). **Intestinal goblet cells and mucins in health and disease: recent insights and progress.** *Curr Gastroenterol Rep* 12: 319-330.

90  Kleerebezem M, van Kranenburg R, Tuinier R, Boels IC, Zoon P, Looijesteijn E, Hugenholtz J et al. (1999). **Exopolysaccharides produced by *Lactococcus lactis*: from genetic engineering to improved rheological properties?** *Antonie Van Leeuwenhoek* 76: 357-365.

91  Kleessen B, Hartmann L and Blaut M (2001). **Oligofructose and long-chain inulin: influence on the gut microbial ecology of rats associated with a human faecal flora.** *Br J Nutr* 86: 291-300.

92  Kornberg HL, Lambourne LT and Sproul AA (2000). **Facilitated diffusion of fructose via the phosphoenolpyruvate/glucose phosphotransferase system of *Escherichia coli*.** *Proc Natl Acad Sci U S A* 97: 1808-1812.

93  Kultz D and Chakravarty D (2001). **Hyperosmolality in the form of elevated NaCl but not urea causes DNA damage in murine kidney cells.** *Proc Natl Acad Sci U S A* 98: 1999-2004.

94  Laemmli UK (1970). **Cleavage of structural proteins during the assembly of the head of bacteriophage T4.** *Nature* 227: 680-685.

95  Lagarde AE and Stoeber FR (1975). **The energy-coupling controlled efflux of 2-keto-3-deoxy-D-gluconate in *Escherichia coli* K 12.** *Eur J Biochem* 55: 343-354.

96  Lancaster JR, Jr. and Hinkle PC (1977). **Studies of the beta-galactoside transporter in inverted membrane vesicles of *Escherichia coli*. II. Symmetrical binding of a dansylgalactoside induced by an electrochemical proton gradient and by lactose efflux.** *J Biol Chem* 252: 7662-7666.

97  Lazim Z and Rowbury RJ (2000). **An extracellular sensor and an extracellular induction component are required for alkali induction**

of alkyl hydroperoxide tolerance in *Escherichia coli*. *J Appl Microbiol* 89: 651-656.

98 Lee C, Lee SM, Mukhopadhyay P, Kim SJ, Lee SC, Ahn WS, Yu MH et al. (2004). **Redox regulation of OxyR requires specific disulfide bond formation involving a rapid kinetic reaction path**. *Nat Struct Mol Biol* 11: 1179-1185.

99 Lee JW, Choi S, Park JH, Vickers CE, Nielsen LK and Lee SY (2010a). **Development of sucrose-utilizing *Escherichia coli* K-12 strain by cloning beta-fructofuranosidases and its application for L-threonine production**. *Appl Microbiol Biotechnol* 88: 905-913.

100 Lee YK and Mazmanian SK (2010b). **Has the microbiota played a critical role in the evolution of the adaptive immune system?** *Science* 330: 1768-1773.

101 Ley RE, Hamady M, Lozupone C, Turnbaugh PJ, Ramey RR, Bircher JS, Schlegel ML et al. (2008). **Evolution of mammals and their gut microbes**. *Science* 320: 1647-1651.

102 Mackie RI and White BA (1997). Gastrointestinal Microbiology. Vol 1 Gastrointestinal Ecosystems and Fermentations. New York, Chapman and Hall Dept. BC.

103 *Maczulak AE, Wolin MJ and Miller TL (1993).* **Amounts of viable anaerobes, methanogens, and bacterial fermentation products in feces of rats fed high-fiber or fiber-free diets**. *Appl Environ Microbiol* 59: 657-662.

104 *Magwedere K and Mukaratirwa S (2008).* **Evaluation of Intestinal pH and Osmolality Levels in Rats (*Rattus norvegicus*) and Chickens (*Gallus gallus*) Experimentally Infected With *Trichinella zimbabwensis***. *Intern J Appl Res Vet Med* 6: 199-174.

105 *Malpica R, Sandoval GR, Rodriguez C, Franco B and Georgellis D (2006).* **Signaling by the arc two-component system provides a link between the redox state of the quinone pool and gene expression**. *Antioxid Redox Signal* 8: 781-795.

106 *Mansson I* (1957). **The intestinal flora in horses with certain skin changes; with special reference to the coliform microbes.** *Acta Pathol Microbiol Scand Suppl* 119: 1-102.

107 *Martinez A and Kolter R* (1997). **Protection of DNA during oxidative stress by the nonspecific DNA-binding protein Dps.** *J Bacteriol* 179: 5188-5194.

108 *Mata-Gilsinger M, Ritzenthaler P and Blanco C* (1983). **Characterization of the operator sites of the *exu* regulon in *Escherichia coli* K-12 by operator-constitutive mutations and repressor titration.** *Genetics* 105: 829-842.

109 *McLaggan D, Naprstek J, Buurman ET and Epstein W* (1994). **Interdependence of K+ and glutamate accumulation during osmotic adaptation of *Escherichia coli*.** *J Biol Chem* 269: 1911-1917.

110 *McNeil NI* (1984). **The contribution of the large intestine to energy supplies in man.** *Am J Clin Nutr* 39: 338-342.

111 *Meryman HT* (1971). **Osmotic stress as a mechanism of freezing injury.** *Cryobiology* 8: 489-500.

112 *Miller CG* (1975). **Peptidases and proteases of *Escherichia coli* and *Salmonella typhimurium*.** *Annu Rev Microbiol* 29: 485-504.

113 *Moazed D and Noller HF* (1987). **Chloramphenicol, erythromycin, carbomycin and vernamycin B protect overlapping sites in the peptidyl transferase region of 23S ribosomal RNA.** *Biochimie* 69: 879-884.

114 *Möller AK, Leatham MP, Conway T, Nuijten PJ, de Haan LA, Krogfelt KA and Cohen PS* (2003). **An *Escherichia coli* MG1655 lipopolysaccharide deep-rough core mutant grows and survives in mouse cecal mucus but fails to colonize the mouse large intestine.** *Infect Immun* 71: 2142-2152.

115 *Moon TS, Yoon SH, Tsang Mui Ching MJ, Lanza AM and Prather KL* (2009). **Enzymatic assay of D-glucuronate using uronate dehydrogenase.** *Anal Biochem* 392: 183-185.

116 *Moore WE and Holdeman LV* (1974). **Human fecal flora: the normal flora of 20 Japanese-Hawaiians.** *Appl Microbiol* 27: 961-979.

117 *Muir M, Williams L and Ferenci T* (1985). **Influence of transport energization on the growth yield of Escherichia coli.** *J Bacteriol* 163: 1237-1242.

118 *Myhal ML, Laux DC and Cohen PS* (1982). **Relative colonizing abilities of human fecal and K 12 strains of Escherichia coli in the large intestines of streptomycin-treated mice.** *Eur J Clin Microbiol* 1: 186-192.

119 *Ochsner UA, Vasil ML, Alsabbagh E, Parvatiyar K and Hassett DJ* (2000). **Role of the Pseudomonas aeruginosa oxyR-recG operon in oxidative stress defense and DNA repair: OxyR-dependent regulation of katB-ankB, ahpB, and ahpC-ahpF.** *J Bacteriol* 182: 4533-4544.

120 *Patrone V, Ferrari S, Lizier M, Lucchini F, Minuti A, Tondelli B, Trevisi E et al.* (2012). **Short-term modifications in the distal gut microbiota of weaning mice induced by a high-fat diet.** *Microbiology* 158: 983-992.

121 *Peekhaus N and Conway T* (1998). **What's for dinner?: Entner-Doudoroff metabolism in Escherichia coli.** *J Bacteriol* 180: 3495-3502.

122 *Pesti L* (1963). **Qualitative and Quantitative Examination of the Intestinal Bacterium Flora of Healthy Pigs.** *Zentralbl Bakteriol Orig* 189: 282-293.

123 *Pikis A, Hess S, Arnold I, Erni B and Thompson J* (2006). **Genetic requirements for growth of Escherichia coli K12 on methyl-alpha-D-glucopyranoside and the five alpha-D-glucosyl-D-fructose isomers of sucrose.** *J Biol Chem* 281: 17900-17908.

124 *Portalier R, Robert-Baudouy J and Stoeber F* (1980). **Regulation of Escherichia coli K-12 hexuronate system genes: exu regulon.** *J Bacteriol* 143: 1095-1107.

125 *Portalier RC and Stoeber FR* (1972). **[D-mannonate: NAD oxidoreductase from *Escherichia coli* K12. Purification, properties and specificity]**. *Eur J Biochem* 26: 290-300.

126 *Prodromou C, Artymiuk PJ and Guest JR* (1992). **The aconitase of *Escherichia coli*. Nucleotide sequence of the aconitase gene and amino acid sequence similarity with mitochondrial aconitases, the iron-responsive-element-binding protein and isopropylmalate isomerases**. *Eur J Biochem* 204: 599-609.

127 *Qin J, Li R, Raes J, Arumugam M, Burgdorf KS, Manichanh C, Nielsen T et al.* (2010). **A human gut microbial gene catalogue established by metagenomic sequencing**. *Nature* 464: 59-65.

128 *Rabilloud T, Strub JM, Luche S, van Dorsselaer A and Lunardi J* (2001). **A comparison between Sypro Ruby and ruthenium II tris (bathophenanthroline disulfonate) as fluorescent stains for protein detection in gels**. *Proteomics* 1: 699-704.

129 *Richey DP and Lin EC* (1972). **Importance of facilitated diffusion for effective utilization of glycerol by *Escherichia coli***. *J Bacteriol* 112: 784-790.

130 *Ritzenthaler P, Mata-Gilsinger M and Stoeber F* (1981). **Molecular cloning of *Escherichia coli* K-12 hexuronate system genes: *exu* region**. *J Bacteriol* 145: 181-190.

131 *Robert-Baudouy JM and Stoeber FR* (1973). **Purification and properties of D-mannonate hydrolyase from *Escherichia coli* K12**. *Biochim Biophys Acta* 309: 473-485.

132 *Rocha ER, Owens G, Jr. and Smith CJ* (2000). **The redox-sensitive transcriptional activator OxyR regulates the peroxide response regulon in the obligate anaerobe *Bacteroides fragilis***. *J Bacteriol* 182: 5059-5069.

133 *Rocha ER, Selby T, Coleman JP and Smith CJ* (1996). **Oxidative stress response in an anaerobe, *Bacteroides fragilis*: a role for catalase in protection against hydrogen peroxide**. *J Bacteriol* 178: 6895-6903.

134 *Rodionov DA, Gelfand MS and Hugouvieux-Cotte-Pattat N* (2004). **Comparative genomics of the KdgR regulon in *Erwinia chrysanthemi* 3937 and other gamma-proteobacteria**. *Microbiology* 150: 3571-3590.

135 *Rodionov DA, Mironov AA, Rakhmaninova AB and Gelfand MS* (2000). **Transcriptional regulation of transport and utilization systems for hexuronides, hexuronates and hexonates in gamma purple bacteria**. *Mol Microbiol* 38: 673-683.

136 *Round JL and Mazmanian SK* (2009). **The gut microbiota shapes intestinal immune responses during health and disease**. *Nat Rev Immunol* 9: 313-323.

137 *Sakamoto N, Kotre AM and Savageau MA* (1975). **Glutamate dehydrogenase from *Escherichia coli*: purification and properties**. *J Bacteriol* 124: 775-783.

138 *Salmon KA, Hung SP, Steffen NR, Krupp R, Baldi P, Hatfield GW and Gunsalus RP* (2005). **Global gene expression profiling in *Escherichia coli* K12: effects of oxygen availability and ArcA**. *J Biol Chem* 280: 15084-15096.

139 *Savage DC* (1986). **Gastrointestinal microflora in mammalian nutrition**. *Annu Rev Nutr* 6: 155-178.

140 *Schmid K, Schupfner M and Schmitt R* (1982). **Plasmid-mediated uptake and metabolism of sucrose by *Escherichia coli* K-12**. *J Bacteriol* 151: 68-76.

141 *Schumann S, Alpert C, Engst W, Loh G and Blaut M* (2012). **Dextran sodium sulfate-induced inflammation alters the expression of proteins by intestinal *Escherichia coli* strains in a gnotobiotic mouse model**. *Appl Environ Microbiol* 78: 1513-1522.

142 *Simmering R, Kleessen B and Blaut M* (1999). **Quantification of the flavonoid-degrading bacterium *Eubacterium ramulus* in human fecal samples with a species-specific oligonucleotide hybridization probe**. *Appl Environ Microbiol* 65: 3705-3709.

143 *Slack E, Hapfelmeier S, Stecher B, Velykoredko Y, Stoel M, Lawson MA, Geuking MB et al.* (2009). **Innate and adaptive immunity**

cooperate flexibly to maintain host-microbiota mutualism. *Science* 325: 617-620.

144 Smith RA, Porteous CM, Coulter CV and Murphy MP (1999). **Selective targeting of an antioxidant to mitochondria.** *Eur J Biochem* 263: 709-716.

145 Stevenson G, Andrianopoulos K, Hobbs M and Reeves PR (1996). **Organization of the *Escherichia coli* K-12 gene cluster responsible for production of the extracellular polysaccharide colanic acid.** *J Bacteriol* 178: 4885-4893.

146 Stock JB, Rauch B and Roseman S (1977). **Periplasmic space in *Salmonella typhimurium* and *Escherichia coli*.** *J Biol Chem* 252: 7850-7861.

147 Stojiljkovic I, Baumler AJ and Hantke K (1994). **Fur regulon in gram-negative bacteria. Identification and characterization of new iron-regulated *Escherichia coli* genes by a fur titration assay.** *J Mol Biol* 236: 531-545.

148 Storz G and Imlay JA (1999). **Oxidative stress.** *Curr Opin Microbiol* 2: 188-194.

149 Storz G, Tartaglia LA and Ames BN (1990a). **The OxyR regulon.** *Antonie Van Leeuwenhoek* 58: 157-161.

150 Storz G, Tartaglia LA and Ames BN (1990b). **Transcriptional regulator of oxidative stress-inducible genes: direct activation by oxidation.** *Science* 248: 189-194.

151 Storz G, Tartaglia LA, Farr SB and Ames BN (1990c). **Bacterial defenses against oxidative stress.** *Trends Genet* 6: 363-368.

152 Takeda K, Kato M, Wu J, Iwashita T, Suzuki H, Takahashi M and Nakashima I (2001). **Osmotic stress-mediated activation of RET kinases involves intracellular disulfide-bonded dimer formation.** *Antioxid Redox Signal* 3: 473-482.

153 Tao K, Makino K, Yonei S, Nakata A and Shinagawa H (1991). **Purification and characterization of the *Escherichia coli* OxyR**

protein, the positive regulator for a hydrogen peroxide-inducible regulon. *J Biochem* 109: 262-266.

154 Tap J, Mondot S, Levenez F, Pelletier E, Caron C, Furet JP, Ugarte E et al. (2009). **Towards the human intestinal microbiota phylogenetic core**. *Environ Microbiol* 11: 2574-2584.

155 Thoden JB, Raushel FM, Benning MM, Rayment I and Holden HM (1999). **The structure of carbamoyl phosphate synthetase determined to 2.1 A resolution**. *Acta Crystallogr D Biol Crystallogr* 55: 8-24.

156 Toledano MB, Kullik I, Trinh F, Baird PT, Schneider TD and Storz G (1994). **Redox-dependent shift of OxyR-DNA contacts along an extended DNA-binding site: a mechanism for differential promoter selection**. *Cell* 78: 897-909.

157 Traore DA, El Ghazouani A, Ilango S, Dupuy J, Jacquamet L, Ferrer JL, Caux-Thang C et al. (2006). **Crystal structure of the apo-PerR-Zn protein from *Bacillus subtilis***. *Mol Microbiol* 61: 1211-1219.

158 Troelsen JT, Olsen J, Noren O and Sjostrom H (1992). **A novel intestinal trans-factor (NF-LPH1) interacts with the lactase-phlorizin hydrolase promoter and co-varies with the enzymatic activity**. *J Biol Chem* 267: 20407-20411.

159 Turnbaugh PJ, Hamady M, Yatsunenko T, Cantarel BL, Duncan A, Ley RE, Sogin ML et al. (2009a). **A core gut microbiome in obese and lean twins**. *Nature* 457: 480-484.

160 Turnbaugh PJ, Quince C, Faith JJ, McHardy AC, Yatsunenko T, Niazi F, Affourtit J et al. (2010). **Organismal, genetic, and transcriptional variation in the deeply sequenced gut microbiomes of identical twins**. *Proc Natl Acad Sci U S A* 107: 7503-7508.

161 Turnbaugh PJ, Ridaura VK, Faith JJ, Rey FE, Knight R and Gordon JI (2009b). **The effect of diet on the human gut microbiome: a metagenomic analysis in humanized gnotobiotic mice**. *Sci Transl Med* 1: 6ra14.

162 van der Waaij D (1989). **The ecology of the human intestine and its consequences for overgrowth by pathogens such as *Clostridium difficile*.** *Annu Rev Microbiol* 43: 69-87.

163 Vogel-Scheel J, Alpert C, Engst W, Loh G and Blaut M (2010). **Requirement of purine and pyrimidine synthesis for colonization of the mouse intestine by *Escherichia coli*.** *Appl Environ Microbiol* 76: 5181-5187.

164 Weber A, Kogl SA and Jung K (2006). **Time-dependent proteome alterations under osmotic stress during aerobic and anaerobic growth in *Escherichia coli*.** *J Bacteriol* 188: 7165-7175.

165 Weurding RE, Veldman A, Veen WA, van der Aar PJ and Verstegen MW (2001). **Starch digestion rate in the small intestine of broiler chickens differs among feedstuffs.** *J Nutr* 131: 2329-2335.

166 Wilfart A, Montagne L, Simmins PH, van Milgen J and Noblet J (2007). **Sites of nutrient digestion in growing pigs: effect of dietary fiber.** *J Anim Sci* 85: 976-983.

167 Wolf SG, Frenkiel D, Arad T, Finkel SE, Kolter R and Minsky A (1999). **DNA protection by stress-induced biocrystallization.** *Nature* 400: 83-85.

168 Wostmann BS, Larkin C, Moriarty A and Bruckner-Kardoss E (1983). **Dietary intake, energy metabolism, and excretory losses of adult male germfree Wistar rats.** *Lab Anim Sci* 33: 46-50.

169 Wu GD, Chen J, Hoffmann C, Bittinger K, Chen YY, Keilbaugh SA, Bewtra M et al. (2011). **Linking long-term dietary patterns with gut microbial enterotypes.** *Science* 334: 105-108.

170 Xiao M, Xu P, Zhao J, Wang Z, Zuo F, Zhang J, Ren F et al. (2011). **Oxidative stress-related responses of *Bifidobacterium longum* subsp. *longum* BBMN68 at the proteomic level after exposure to oxygen.** *Microbiology* 157: 1573-1588.

171 Yamamoto Y and Gaynor RB (2001). **Therapeutic potential of inhibition of the NF-kappaB pathway in the treatment of inflammation and cancer.** *J Clin Invest* 107: 135-142.

172 Yan D, Ikeda TP, Shauger AE and Kustu S (1996). **Glutamate is required to maintain the steady-state potassium pool in *Salmonella typhimurium*.** *Proc Natl Acad Sci U S A* 93: 6527-6531.

173 Yatsunenko T, Rey FE, Manary MJ, Trehan I, Dominguez-Bello MG, Contreras M, Magris M et al. (2012). **Human gut microbiome viewed across age and geography.** *Nature* 486: 222-227.

174 Zheng M, Aslund F and Storz G (1998). **Activation of the OxyR transcription factor by reversible disulfide bond formation.** *Science* 279: 1718-1721.

175 Zheng M, Wang X, Doan B, Lewis KA, Schneider TD and Storz G (2001a). **Computation-directed identification of OxyR DNA binding sites in *Escherichia coli*.** *J Bacteriol* 183: 4571-4579.

176 Zheng M, Wang X, Templeton LJ, Smulski DR, LaRossa RA and Storz G (2001b). **DNA microarray-mediated transcriptional profiling of the *Escherichia coli* response to hydrogen peroxide.** *J Bacteriol* 183: 4562-4570.

SUPPLEMENTAL MATERIAL

## APPENDIX II:   SUPPLEMENTAL MATERIAL

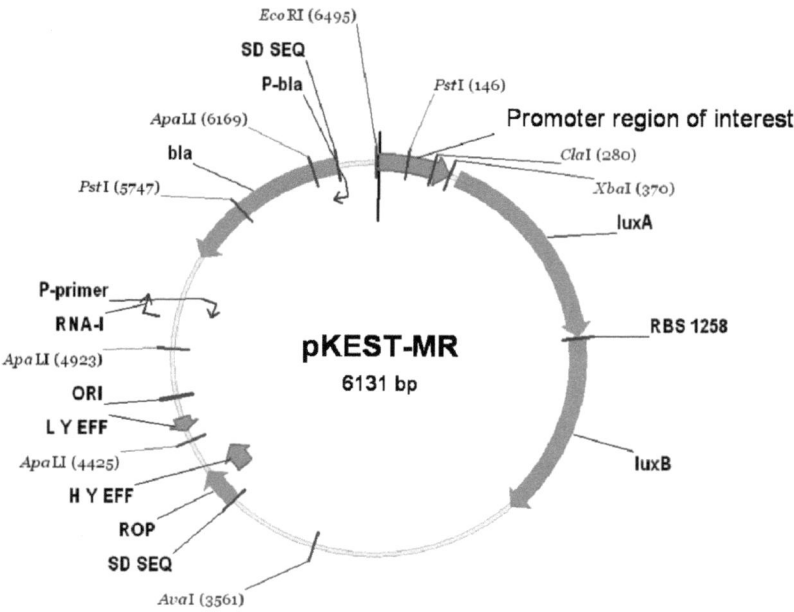

**Figure I.** **Schematic representation of pKEST-MR.** The promoter region of interest is cloned in front of the bacterial luciferase genes *luxAB* and therefore regulates the expression of these genes.

# SUPPLEMENTAL MATERIAL

A
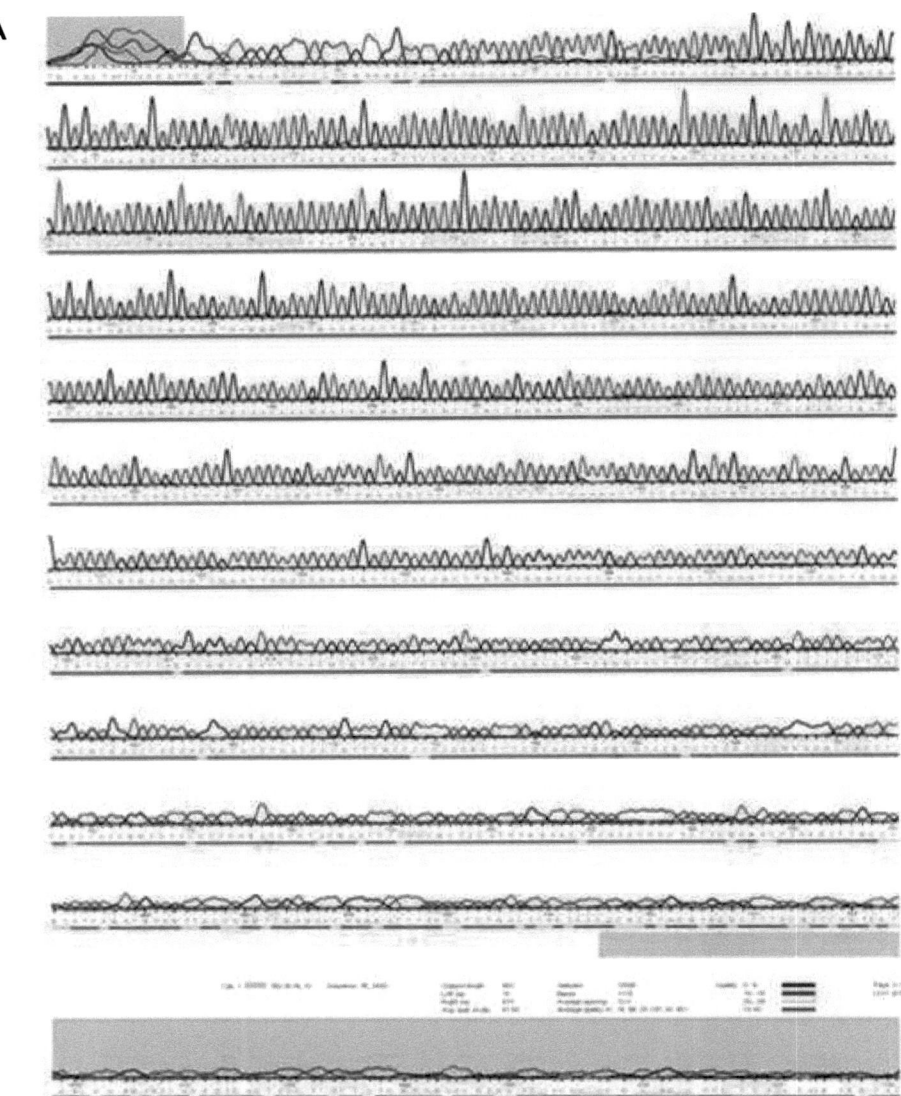

SUPPLEMENTAL MATERIAL

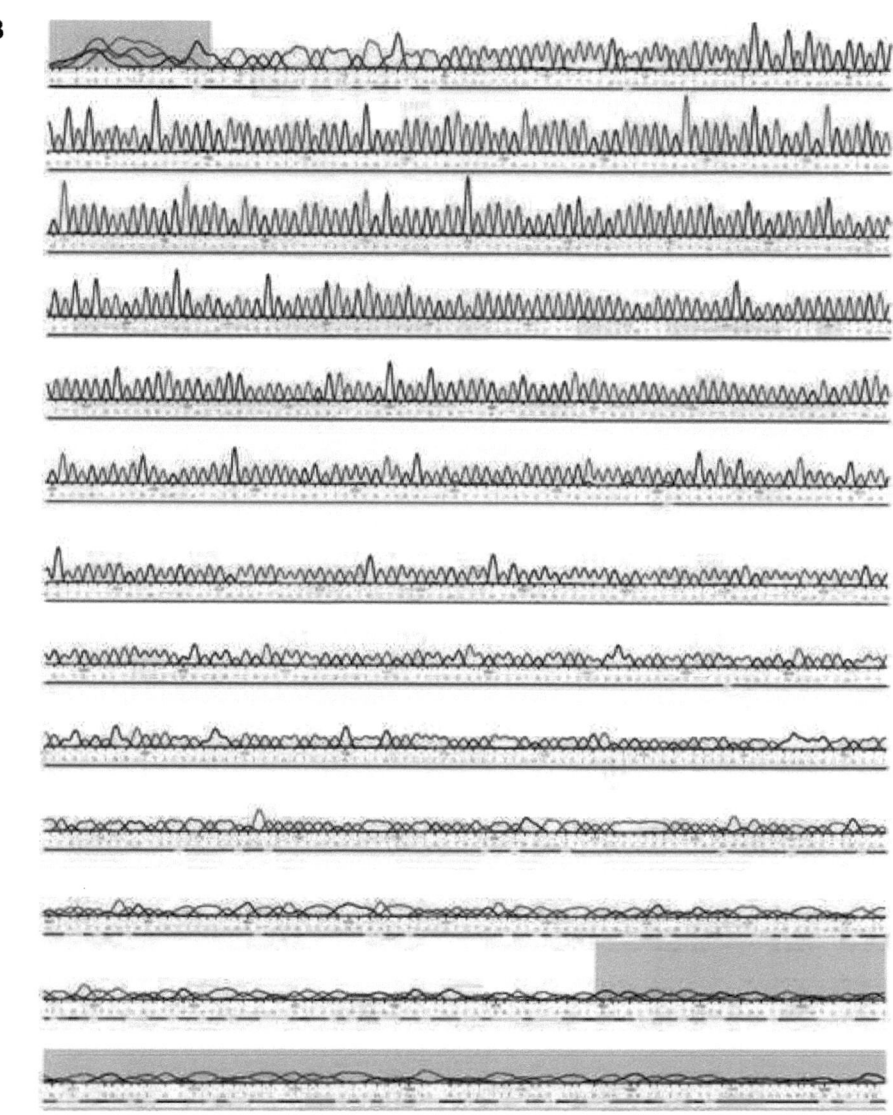

**Figure II.** **Representative 16S rRNA gene sequence of *E. coli*** recovered from caecum of mice fed the starch or the lactose diet. DNA was amplified using 1492-r primer (5'-TAC CTT GTT ACG ACT T-3') [KAGEYAMA, 1999]. Animal 8, starch diet.

# SUPPLEMENTAL MATERIAL

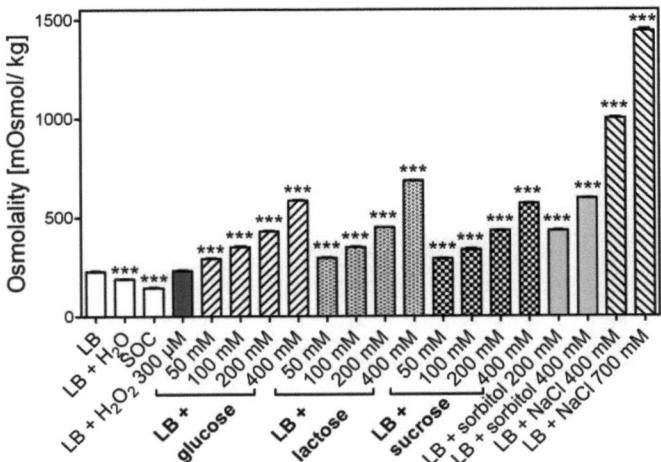

**Figure VI.** **Osmolality of the various media used for the luciferase reporter gene assays** addressing the *ahpCF* and *dps* promoters. LB medium, SOC medium and LB medium supplemented with water, $H_2O_2$, NaCl or different types and concentrations of carbohydrates were investigated. Data are expressed as means, n = 3. Differences between treatment groups were calculated by 1way-AOV and Dunnett's Multiple Comparison Test. **, $P < 0.01$; ***, $P < 0.001$.

# SUPPLEMENTAL MATERIAL

**Table I.** Primers used for generation of luciferase reporter gene constructs, deletion mutants, complementing plasmids and pGEM-T-Easy vectors.

| Code | Amplified region | Sequence (5' → 3') | Annealing temp. | Reference |
|---|---|---|---|---|
| **Generation and verification of luciferase reporter gene constructs** | | | | |
| Primer-ahpCFp | Promoter region of ahpCF | CGGAATTCTCAGTCAGTGCAAAAGTCGAG [a]<br>CGTCTAGAAGGACATCTATACTTCCTCCG [b] | 60 °C | This work |
| Primer-dpsp | Promoter region of dps | GCTCTAGATAAAGCAGATTG [a]<br>GCGAATTCTTGAATCTTTATTAGT [b] | 48 °C | This work |
| Primer-kduIp | Promoter region of ahpCF | CGGAATTCGCCAGGGTGTGGCATTGAC [a]<br>GCTCTAGAGTGCGCACTGTGGATGCTCT [b] | 60 °C | This work |
| Primer-kduDp | Promoter region of kduD | CGGAATTCTAACCTCACCAGTAACCGTCG [a]<br>GCTCTAGAACGACCGCAACTTTACCTTC [b] | 60 °C | This work |
| Primer-uxaCAp | Promoter region of uxaCA | CGTCTAGACAATTTCCAGAGTCCGA [a]<br>CGGAATTCGAAGATGTTAGTTACG [b] | 50 °C | This work |
| Primer-uxaBp | Promoter region of uxaB | CGTCTAGAATGGGTTCCCTTCTGA [a]<br>CGGAATTCATACGTGTCTGTATC [b] | 50 °C | This work |
| Primer-uxuABp | Promoter region of uxuAB | CGGAATTCGGTCAACCATTGTTGCGATG [a]<br>CGTCTAGAGACCGCAGCCGTGGCG [b] | 62 °C | This work |
| pKESTMR control | Integration site of pKESTMR | AAAGTGCCACCTGACGT [a]<br>GGGTTGGTATGTAAGCAA [b] | 46 °C | This work |
| **Generation and verification of *E. coli* deletion mutants** | | | | |
| PrimerΔ ahpCF | Flanking region of ahpCF | AAAAATTGGTTACCTTACATCTCATCGAAAACACGG AGGAAGTATAGATGATTCCGGGGATCCGTCGACC [a]<br>AAGCAATTGCAGGTGAATCTTACTTCTTCTTATGCAG TTTTGGTGCGAATTGTAGGCTGGAGCTGCTTCG [b] | 55 °C | BABA, 2006 |
| PrimerΔ oxyR | Flanking region of oxyR | CTATTCTACCTATCGCCATGAACTATCGTGGCGATG GAGGATGGATAATGATTCCGGGGATCCGTCGACC [a]<br>AAGCCTATCGGGTAGCTGCGTTAAACGGTTTAAACC GCCTGTTTTAAAACTGTAGGCTGGAGCTGCTTCG [b] | 55 °C | BABA, 2006 |
| PrimerΔ kduID | Flanking region of kduID | CACTATCGTTTTCTATTTTCACGCTTCACTGATTATC GGAGGTTGATGTGATTCCGGGGATCCGTCGACC [a]<br>GGCAGGGTCATAAAAGTAAGAAGAATGAATTAACGC GCCAGCCAACCGCCTGTAGGCTGGAGCTGCTTCG [b] | 65 °C | BABA, 2006 |
| K2 | Kanamycin-cassette of pKD13 | GCAGTTCATTCAGGGCACCG [b] | 54 °C | DATSENKO, 2000 |
| Kt | | CGGCCACAGTCGATGAATCC [a] | 54 °C | |
| ahpCF-control | Up-/downstream of ahpCF | CGCATTAGCCGAATCGGC [a]<br>ATAAGTATCCCGCCCTGCCC [b] | 55 °C | This work |

126

## SUPPLEMENTAL MATERIAL

| Code | Amplified region | Sequence (5' → 3') | Annealing temp. | Reference |
|---|---|---|---|---|
| oxyR-control | Up-/downstream of oxyR | GCTGCAATCGTGCCTCGACA [a]<br>TCGTCGGCATGAACGTGGG [b] | 55 °C | This work |
| kduID control | Up-/downstream of kduID | ATATTCGTGATCGACACTGCACTT [a]<br>CACGCAGGTGTCAGGTCGGAA [b] | 54 °C | This work |
| **Generation and verification of complementing pSU19 plasmids** | | | | |
| ahpCF-compl | ahpCF | GCAAGCTTGTCGAGTAAAAGGCATAACCT [a]<br>TAGGATCCAAAGCCGCCAGGTTTGA [b] | 57 °C | This work |
| oxyR-compl | oxyR | GCAAGCTTGTGCCGCTCCGTTTCTGTGA [a]<br>GCGGATCCAACTACCCGACGATGGCGGAA [b] | 66 °C | This work |
| kduID-compl | kduID | CGGAATTCGCTCTGCATTTCCTCCTTAC [a]<br>TACTGCAGCCATGCGGCAGGGTCATAAA [b] | 63 °C | This work |
| pSU19-control | Up-/downstream of MCS | CCAGGCTTTACACTTTATGC [a]<br>AGGCTGCGCAACTGTTG [b] | 50 °C | This work |
| **Generation and verification of pGEM-T Easy vectors** | | | | |
| pGEM-kduI | kduI | GATTATCGGAGGTTGATGTGGA [a]<br>CGTTTATGCCCACAACTAGCG [b] | 52 °C | This work |
| pGEM-kduD | kduD | GCTAGTTGTGGGCATAAACG [a]<br>GTAAGAAGAATGAATTAACGCGCC [b] | 52 °C | This work |
| pGEM-kduID | kduID | GATTATCGGAGGTTGATGTGGA [a]<br>GTAAGAAGAATGAATTAACGCGCC [b] | 52 °C | This work |

[a] forward primer, [b] reverse primer

**Table II.** Identified proteins of *E. coli* obtained from small intestine and caecum of mice fed the starch, the lactose or the casein diet with differential expression factors of ≥ 2 [a].

| Swiss-Prot accession no. | Gene | Protein description | Fold change [b] | | | |
|---|---|---|---|---|---|---|
| | | | Lactose vs. Starch diet | | Casein vs. Starch diet | |
| | | | SI | Cae | SI | Cae |
| **Galactose metabolic processes** | | | | | | |
| P0A9C3 | galM | Galactose mutarotase | | 4.3 | | |
| P0A6T3 | galK | Galactokinase | 2.3 | 6.6 | | |
| P09148 | galT | Galactose-1-phosphate uridylyltransferase | 2.2[c]<br>2.6 | 7.8 | | |
| P09147 | galE | UDP-glucose 4-epimerase | | 3.6 | | |
| P0AEE5 | dgaL | D-galactose-binding periplasmic protein | -2.6 | | | |

| UniProt | Gene | Description | | | | | |
|---|---|---|---|---|---|---|---|
| **Utilisation of further carbohydrates / -acids** | | | | | | | |
| P39346 | idnD | L-idonate 5-dehydrogenase | | | | | 2.9 |
| P0A9P9 | idnO | Gluconate 5-dehydrogenase | | | | 2.2$^c$ 5.6 | |
| P45541 | frlC | Fructoselysine 3-epimerase | | | -4.8 | | |
| P39300 | ulaG | Probable L-ascorbate-6-phosphate lactonase ulaG | 2.2 | | -4.1 | | -2.7 |
| Q46938 | kduI | Predicted 5-keto 4-deoxyuronate-isomerase | | | | | -8.3 |
| P37769 | kduD | 2-deoxy-D-gluconate 3-dehydrogenase | | | | 2.4 | -4.0 |
| P37647 | kdgK | 2-dehydro-3-deoxygluconokinase | | | | | -2.4 |
| P06720 | agaL | Alpha-galactosidase | -2.5 | | | | |
| P0A6L4 | nanA | N-acetylneuraminate lyase | | | -3.5 | | |
| P52643 | ldhA | D-lactate dehydrogenase | 3.8 | | | | |
| **Carbohydrate metabolism** | | | | | | | |
| P0A9B2 | gapA | Glyceraldehyde-3-phosphate dehydrogenase A | | 2.3$^c$ 3.4 2.9 | | 2.1$^c$ 2.7 3.8 | -2.4 |
| P0A799 | pgk | Phosphoglycerate kinase | 2.1 | | | | |
| P0AD61 | pykF | Pyruvate kinase I | | | | 4.2$^c$ 3.0 | |
| P21599 | pykA | Pyruvate kinase II | | | | | -2.8 |
| P0A955 | eda | KHG/KDPG aldolase | | | | 2.0$^c$ 2.0 2.1 | 3.2$^c$ 2.9 |
| P0A9P0 | lpdA | Dihydrolipoyl dehydrogenase | 3.8 | | | | |
| P22259 | pckA | Phosphoenolpyruvate carboxykinase [ATP] | | | -2.4 | | 2.4 |
| P0A9C9 | glpX | Fructose-1,6-bisphosphatase class 2 | 3.9 | | | | |
| P0A799 | pgk | Phosphoglycerate kinase | 2.1 | | | | |
| **Lipid metabolism** | | | | | | | |
| P0AEK4 | fabI | Enoyl-[acyl-carrier-protein] reductase [NADH] | 2.8 | | | | |
| P0ABD8 | fabE | Biotin carboxyl carrier protein of acetyl-CoA carboxylase | | | -3.0 | | |
| P76015 | dhaK | Dihydroxyacetone-binding subunit dhaK | | | -2.5 | | |
| P0A6F3 | glpK | Glycerol kinase | 2.2 | | | | 3.2 |
| P27830 | frrG | dTDP-glucose 4,6-dehydratase 2 | 2.3 | | | | |
| P37744 | rlmA | Glucose-1-phosphate thymidylyltransferase 1 | | | | 2.3 | |
| **Protein- / amino acid biosynthesis / metabolism** | | | | | | | |
| P0A853 | tnaA | Tryptophanase | 3.9 | | | 3.2 | 2.0$^c$ 2.6 |
| P77454 | glsA1 | Glutaminase 1 | | | -3.0 | | |
| P21165 | pepQ | Xaa-Pro dipeptidase | 3.4 | | | | |
| P0ACC7 | glmU | Glucosamine-1-phosphate N-acetyltransferase | | | | | 3.0 |
| 0AB77 | kbl | 2-amino-3-ketobutyrate coenzyme A ligase | 2.5 | | | | |
| P52643 | ldhA | D-lactate dehydrogenase | 3.8 | | | | |
| P0A825 | glyA | Serine hydroxymethyltransferase | 3.5 | | | 6.2$^c$ 5.1 2.4 3.6 | 2.0 |
| P0A9D8 | dapD | 2,3,4,5-tetrahydropyridine-2,6-dicarboxylate N- | 3.0$^c$ | -3.9$^c$ | | | |

## SUPPLEMENTAL MATERIAL

| | | | | | | |
|---|---|---|---|---|---|---|
| | | succinyltransferase | 3.5 | -2.3 | | |
| P0A6D7 | aroK | Shikimate kinase 1 | | | | -2.2 |
| P60757 | hisG | ATP phosphoribosyltransferase | | -2.2 | | |
| P0ABK5 | cysK | Cysteine synthase A | 2.9 | -8.0 | | |
| P00370 | gdhA | NADP-specific glutamate dehydrogenase | | 2.3 | | |
| P05793 | ilvC | Ketol-acid reductoisomerase | | -3.1 | | -3.4 |
| P11447 | argH | Argininosuccinate lyase | | | | 2.1 |
| P0A6F1 | carA | Carbamoyl-phosphate synthase small chain | 3.2 | 2.9 | | |

| **Cell cycle (translation, transcription, cell division)** | | | | | | |
|---|---|---|---|---|---|---|
| P0A8G6 | wrbA | Flavoprotein wrbA | | | -2.1 | |
| P0CE47 | tufA | Elongation factor Tu 1 | 5.8 | -2.0$^c$ -2.2<br>-3.3 -4.6 | | |
| P0A6P1 | eftS | Elongation factor Ts | | | -2.4 | |
| P0A7V8 | rpsD | 30S ribosomal protein S4 | | | | 2.5 |
| P02358 | rpsF | 30S ribosomal protein S6 | -2.0 | | -2.3 | |
| P0A7R5 | rpsJ | 30S ribosomal protein S10 | | | -2.1 | |
| P60906 | hisS | Histidyl-tRNA synthetase | 2.4 | | | |
| P04805 | gltX | Glutamyl-tRNA synthetase | 2.6 | | | |
| P0AEZ3 | minD | Septum site-determining protein minD | | -3.3 | | |
| P0AGE0 | Ssb | Single-stranded DNA-binding protein | | -2.9 | | |

| **Nucleotide metabolic process** | | | | | | |
|---|---|---|---|---|---|---|
| P77671 | allB | Allantoinase | | -5.0 | | |
| P0A794 | pdxJ | Pyridoxine 5'-phosphate synthase | | | -3.2 | |
| P0ABP8 | deoD | Purine nucleoside phosphorylase deoD-type | | -2.2 | -2.0 | |
| P15639 | purH | Bifunctional purine biosynthesis protein purH | 2.2 | | -3.1 | |
| P0AG18 | purE | Phosphoribosylaminoimidazole carboxylase catalytic subunit | | 4.6 | | |
| P0A7D7 | purC | Phosphoribosylaminoimidazole-succinocarboxamide synthase | | 2.5 | -2.6 | |
| P12758 | udp | Uridine phosphorylase | | -2.2 | | |
| P0A8F0 | upp | Uracil phosphoribosyltransferase | | 2.1 | -2.9 | |
| P0AE22 | aphA | Class B acid phosphatase | | -2.4 | | |
| P69441 | adk | Adenylate kinase | -3.0 | | | |
| P0A786 | pyrB | Aspartate carbamoyltransferase catalytic chain | 3.5$^c$<br>4.0 | 17.0$^c$ 8.6<br>4.5 | | 5.2<br>2.1 |

| **Stress response** | | | | | | |
|---|---|---|---|---|---|---|
| P0A9A9 | fur | Ferric uptake regulation protein | | 3.1 | | |
| P0ABT2 | dps | DNA protection during starvation protein | | 3.2 | | |
| P68066 | grcA | Autonomous glycyl radical cofactor | 2.3 | -8.8 | | |
| P0AFF6 | nusA | Transcription elongation protein nusA | 3.1 | | | |
| P0A6H5 | hslU | ATP-dependent protease ATPase subunit HslU | | | -4.5 | |
| P05055 | pnp | Polyribonucleotide nucleotidyltransferase | | -6.6 | | |
| P0A9D2 | gst | Glutathione S-transferase | | | | -2.7 |
| P35340 | ahpF | Alkyl hydroperoxide reductase subunit F | 3.2 | 2.2 | -3.5 | |
| P0AE08 | ahpC | Alkyl hydroperoxide reductase subunit C | | | -2.4 | -2.1 |

# SUPPLEMENTAL MATERIAL

| Accession | Gene | Protein | | | | |
|---|---|---|---|---|---|---|
| P0A862 | tpx | Thiol peroxidase | | | | 2.0 |
| P0ACE0 | hybC | Hydrogenase-2 large chain | 2.6 | 2.0 | | |
| P38489 | nfnB | Oxygen-insensitive NAD(P)H nitroreductase | 2.2 | | | |
| P39315 | qorB | Quinone oxidoreductase 2 | -4.8 | | | |
| **Transport of ions, sugars, amino acids, peptides** | | | | | | |
| P0A910 | ompA | Outer membrane protein A | 2.3 | | | |
| P06996 | ompC | Outer membrane protein C | 2.0 | | | |
| P02931 | ompF | Outer membrane protein F | 2.1 | -3.3 | | 2.5 |
| P69797 | manX | PTS system mannose-specific EIIAB component | | $-2.1^d$ | -3.0 | -3.2 |
| P0AEX9 | malE | Maltose-binding periplasmic protein | 2.5 | | | |
| P02925 | rbsB | D-ribose-binding periplasmic protein | | -6.1 | | -2.8 |
| P23843 | oppA | Periplasmic oligopeptide-binding protein | 3.2 | | -3.2 | |
| P0A855 | tolB | Protein tolB | 2.0 | 3.0 | | 4.5 |
| P0AEM9 | fliY | Cystine-binding periplasmic protein | | | | 2.5 |
| **Uncharacterised proteins** | | | | | | |
| P75682 | yagE | Uncharacterised protein yagE | 2.1 | | | |
| Q46868 | yqiC | Uncharacterised protein yqiC | | -2.5 | | -2.1 |
| P39173 | yeaD | Putative glucose-6-phosphate 1-epimerase | 2.4 | | | |
| Q46803 | ygeW | Uncharacterised protein ygeW | 2.2 | | | |
| P65807 | ygeY | Uncharacterised protein ygeY | | -2.9 | | |
| P76187 | ydhF | Oxidoreductase ydhF | 2.0 | | | |
| P0A8Y5 | yidA | Phosphatase yidA | | -4.3 | | |
| P39831 | ydfG | NADP-dependent L-serine/L-allo-threonine dehydrogenase ydfG | 2.4 | | | |
| P0AD12 | yeeZ | Protein yeeZ | | -5.9 | | -2.6 |
| P77748 | ydiJ | Uncharacterised protein ydiJ | | -2.3 | | |
| **Others** | | | | | | |
| P0ABB0 | atpA | ATP synthase subunit alpha | $3.1^d$ 3.4 | | | |
| P0ABB4 | atpB | ATP synthase subunit beta | | -4.5 | | -4.7 |
| P24186 | folD | Bifunctional protein folD | 2.6 | | | |
| P09127 | hemX | Putative uroporphyrinogen-III C-methyltransferase | | | | |

[a] Comparison of the proteomes of *E. coli* obtained from the intestines of mice fed the lactose or the casein diet with those of mice fed the starch diet.

[b] Data represent average ratios of results from 20 biological replicates per diet. 2D-DIGE analyses were performed using pooled samples, dependent on the available material. SI casein diet, Cae starch diet: duplicate, SI starch diet, lactose diet: triplicate, Cae lactose diet: quadruplicate, Cae casein diet: quintuplicate, P ≤ 0.05. SI = small intestine, Cae = caecum

[c] If different isoforms were identified for the same protein, each isoform was indicated separately.

**SUPPLEMENTAL MATERIAL**

**Table III.** Specific activities of KduI and KduD calculated for hexuronate concentrations observed after incubation of cell-free extracts of *E. coli* clones overexpressing KduI, KduD or both with 10 mM galacturonate or 10 mM glucuronate at 37°C.

| | Specific activity after 2 h [nmol/min*mg] [a] | | | | |
|---|---|---|---|---|---|
| | *E. coli* JM109 pGEM-T | *E. coli* JM109 pGEM-T-*kduID* | *E. coli* KRX pGEM-T | *E. coli* KRX pGEM-T-*kduD* | *E. coli* KRX pGEM-T-*kduI* |
| **Glucuronate** | 1.8 (1.0:2.4) | 5.2 (2.3:6.1) [b] | 0.7 (-0.1:2.9) | 1.2 (0.3:3.0) | 3.7 (2.5:4.9) [b] |
| **Galacturonate** | 1.6 (0.6:3.3) | 3.1 (2.4:8.6) [b] | 0.8 (-0.2:1.6) | 1.3 (0.2:2.6) | 4.4 (2.6:8.9) [b] |

[a] Data are expressed as medians and minima versus maxima (n = 11-12).
[b] Differences between negative controls and expression of the recombinant proteins were calculated by the U-test. $P < 0.001$.

# yes
# i want morebooks!

Buy your books fast and straightforward online - at one of world's fastest growing online book stores! Environmentally sound due to Print-on-Demand technologies.

## Buy your books online at
## www.get-morebooks.com

Kaufen Sie Ihre Bücher schnell und unkompliziert online – auf einer der am schnellsten wachsenden Buchhandelsplattformen weltweit! Dank Print-On-Demand umwelt- und ressourcenschonend produziert.

## Bücher schneller online kaufen
## www.morebooks.de

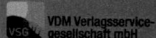 VDM Verlagsservicegesellschaft mbH
Heinrich-Böcking-Str. 6-8   Telefon: +49 681 3720 174   info@vdm-vsg.de
D - 66121 Saarbrücken   Telefax: +49 681 3720 1749   www.vdm-vsg.de

Printed by Books on Demand GmbH, Norderstedt / Germany